NUBAR SARAFYAN

Cycles and Self-Organization

Los Angeles

ANNOTATION

Our idea about the rhythms and cycles of the Nature are of oldest ones. It originated from observations of rhythmic and cyclic alternations in the nature and then became simple and ordinary. Cyclic recurrence of phenomena is observed always and everywhere: in alternations of the celestial bodies trajectories, in the stages of the Earth's evolution, in sedimentation, in oceanic and atmospheric motions, climatic changes, etc. At the same time the science about rhythms, cycles, and their organization, which were rapidly developed in the last century, show that these simplest alternations hide many difficulties which are not clearly understood even now.

This work is devoted to discussion of attractive problems of rhythms and their organization in natural phenomena.

Nubar Sarafyan E.
Cycles and Self-Organization

Includes bibliographic references (p.155-169)

DIGITAL COLOR WORLD.
1636 Wilshire Blvd. Suite 300, Los Angeles, CA 90017 , Tel. 213/353-1460
Fax: 213/3531458   BBS: 213/ 353-4726

Printed in the United States of America

To
My Wife Marine
My Daughters Hripsime and Armine
My Mother Hripsime

Acknowledgments

During the writing this book it has been my good fortune to have the active guidance of many talented people of various scientific fields, whose knowledge and skills have significantly influenced me. I am indebted to those people, who have shared their time and experience to help me prepare this book.

I am deeply grateful first of all to a number of my friends and researchers at the Institute of Geology who either evaluated the separate chapters or reviewed parts of this book. Along with Drs. H. Guyumjyan, B. Meliksetyan, A. Sadoyan, my appreciation is extended to Dr. Pronin (Institute of Geology, Moskow) for criticism and encouragement. I would like to thank to Drs. A. Vehuny, G. Chilingaryan (USC), and R. Ingersoll (UCLA) for many valuable discussions and assistance. My thanks go to Dr. N. Vassoevich, (Moskow State University), who first introduced me to sediment cycles and provided many of the ideas used in their presentation. He inspired the first attempts to write this book. His landmark research and findings on sediment cycles have provided the basis for much of the research currently being done on cycles. My appreciation is also extended to Dr. A.Arakelyan, who first acquainted me with spiral and concentric structures in the lithosphere and offered to examine their genesis and distributions. It was unforgettable the cordial atmosphere, valuable discussions, and conversations regarding diverse points, opinions and comment during the "Mathematical Methods of Recurrence Analyses in Geology" conferences organized regularly by Professor Affanasev (Moskow) and his group during the years 1980 to 1990.

I would like to begin by thanking my friends and investigators from the Institute of Mechanics who offered suggestions and comments.First I would like to mention Dr.L.Agalovyan who first concentrated my attention on the multistage processing in the non-homogeneous solid media, and encouraged me to continue the project. My discussions and conversations with Drs. R. Hovakimyan and S. Zargaryan (University of California, San Diego) regarding this and many other, related mechanics of solid bodies questions were valuable. The suggestions and comments offered by Dr. S. Grigoryan (Institute of Mechanics, Moskow State University) were inestimable and my discussions with Dr. L. Abrahamyan were useful.

I am also deeply grateful to the following people who either evaluated the separate chapters or reviewed parts of this book related to the astronomic and astrophysics issues. First of all I want to thank Drs. B. Panadjyan (Byurakan Observatory) and K. Papoyan (Institute of Physics) who read the first drafts of this material and provided the criticism and encouragement necessary to continue the project. Each continued to be of assistance throughout the project by answering what must have seemed an endless stream of questions about cycles. I want to pay special tribute to the discussions of some astrophysical issues with Dr. V.Ambartsumyan (Byurakan Observatory) Dr. G. Gurzadyan (Garni Cosmic Research Institute) and Dr. G. Sahagyan (Yerevan State University); their comments and suggestions have been significant. My appreciation is extended also to Professor Milan Mjic (California State University) for his criticism and willingness.

I gratefully acknowledge the discussions and assistance of many people, regarding the biological concepts and confirmations of this project. This area is one of the most important fields of investigation for the understanding of some concepts of other scientific fields. After I obtained my M.A. from Yerevan State University, I intensively started to examine biological theories. So, it was not accidental, that my Ph.D. in applied mathematics was dedicated to multilevel control systems,

having in its base the functional and structural features of the central nervous system of living beings.
I would like to express my thanks to many people whose in one way or another contributed to this project though it is impossible to mention them all. However among these people I want to mention Professor L.Gambaryan, (Institute of Experimental Biology), and especially Dr. Y.Svirejev (Institute of Biophysics, Moskow), Dr. E .Asratyan (Institute of Physiology, Moskow). The discussions and conversations with these people have been good background for me. The criticism and suggestions offered by Dr. R.Frysinger (Brain Institute, UCLA) and Dr. M.Arbib (University of Southern California) regarding brain rhythms, were extremely useful.

Our appreciation will be incomplete without mentioning researchers working in the fields of investigation of the hydrosphere and atmosphere. I am deeply thankful these people, who by their grave works and cheerful observations infinitely helped me to form and accomplish ideas and presentations suggested in this book; along with E.Lorenz, E.Palmen, A.Oort, I.Maksimov, K.Fedorov, A.Sarkisyan, S. and many other proficient people, whose enumeration is impossible.

Professor B. Dauzat was the first editor to see this book. He looked at the early manuscript with a critical eye for grammar, punctuation, and sentence structure. Dr.V. Vardanyan and Mr. B. Wright did corrections regarding to terminology and expressions. I would like to thank to all of them.

I must also thank my wife Maria for her support. Writing is a lonely pursuit, because one must be alone with one's thoughts for many hours of the day. I thing her for giving me the time to read, reflect, and write without feeling that I was neglecting her too much. I also thank her and my daughters Hripsime and Armine for the unselfish love, endless patience and quiet understanding that allowed me to devote such a large part of my life to my career. Truly, without their willingness to assume many of the tasks necessary to produce a book and their sacrifices that allowed me the time needed to revise it, this book would never have been completed.

Finally, I must thank NGSR Foundation, whose financial support made it possible to accomplish this, long expected, project.

Contents

Chapter VI

PREFACE

In the opinion of certain researchers, repetitions and rhythms of natural phenomena are unquestionable and obvious, and in the opinion of others, they average and indistinctly pronounced. The forgers have trend to see the basis of the world structure in rhythms, the other state that rhythm is not beyond oscillations of random series. In order to characterize a time history of phenomena periodic and cyclic concepts are often used. According to some researchers, there is no difference among these three concepts, and these laws are superior to other nature's laws. Others attribute rhythm to individual features of observed phenomena, e.g. current regularity, symmetry, duration. Or, they use these concepts only for changes of particular magnitudes. And finally, there is an opinion according to which the concept of rhythm is related to time intervals of deviation from and return to initial states of systems (especially to restoration processes). Rhythmic explanation of system's behavior regular change is convenient in the sense that it focuses attention on phenomenon's time and spatial specific features only. It also introduces a natural classification into interactions and scales of phenomena (bodies) and gives some hopes to clarify a significance of cycle structure in the problems of drift systems' stability.

The point we wish to make in the book is that in nature; atmosphere, hydrosphere, lithosphere in the Earth, in solar system we continually observe energy's transformation with the substance multistage cyclic circulation. Extending this concept to the biogenesis and the living systems, particularly to the human being to modern life, to economic, industrial, scientific spheres, we again see, that we are continually surrounded by substance, information multistage cyclic circulation and decision processes. Arising in a multitude of diverse fields these embedded, hierarchic systems share the same common property: energy transformation is accomplished by multistage cyclic circulation of the substance, information.

Naturally, in one book it is unrealizable to involve all scientific and methodological problems, arising with connection to the comprehension of multistage cyclic changes. Therefore, we will briefly consider only some of them. Taking into account in the first, that the work is devoted the relatively wide round's readers-physics, geologists, astronomers, by various extent familiar with cycles' different aspects, and the second, that cyclic organization and interactions are elucidated, in the literature comparatively weakly, four chapters were devoted to the cycles' features in the astronomy, lithosphere, hydrosphere, atmosphere. In the exposition we will restrict ourselves more, and will evade from definite category cycles. For example, associated with the vapor power cycles, such as Carnot, Ranrine, binary, and also, economical and stock market wide spectrum cycles totally do not analyzed. Biological and ecological rhythms were briefly considered separately, more early. By that sense, the book is not cycles observation, or an attempt of their classification and edges' precision. The cycles' joint examination is stipulated by three reasons. One, to check whether is the talk subject in the each area is about the same qualities and quantities? Two, to elucidate each group's spatial and temporarily structures, organization, and the origin possible mechanisms. Three, to use the one as a model, concept for survey the another, on the base of method analogy. For example, atmosphere cycles consider as a model in the studies hydrosphere cycles, the hydrosphere cycles-as a concept at the research geophysical and geological cycles. This is the only way to make a judgment about structures and interactions of galactic cycles and about the presence more large-scales' structures or more slowly flowing cosmic cycles[1]. In other words, the accessible cycles are used, as a tool in analyses of less accessible and detectable cycles. It is also the only way to obtain worthwhile information about spatial-temporal features of very small and quick altering cycles. The reason of such conduct is: the information abundance and diversity as well as scientific stereotypes often prevent to reveal and to interpret non standard, new laws. In all of our considerations involving cycles self-organization, we implicitly observe that the cycles are extermal reactions, trajectories, and so, the corresponding dynamic systems obligatory also are submitted to an extreme principle.

Our main purpose is; to study specific features of different order rhythms and cycles, to attempt to establish their general properties, the principles and causes of their formation and conservation. Such approach makes possible to understand a role and structure of medium more deeply and correctly, laws of its evolution, and finally to reveal them cooperatively as a consequence of one following from another.

Attributing of rhythms only to one explanation, i.e. to energy conversations (metabolism) as far as convenient and natural it is, is not enough to reveal and understand all properties and links of rhythms. Thus, α, β, γ, δ and other rhythms of brain are related to electric potentials and considering their basis only as concentration oscillations is not enough for explanation of many of specific features of their forming and conservation. Apparently there are other mechanisms equivalent to energy metabolism and which are more perfect in high-frequency domain, in problems of reception, processing and transmission of signals and in establishing more efficient equilibrium.

Due to coincidence of circumstances for various frequency rhythms it is essential to reveal and to solve various problems which, as a rule, are to be made by different methods and with different depth, hence joint consideration will allow to complete their dark sides, to revise states of primary and auxiliary rhythms, their substantial and non-substantial parameters, which are difficult to perceive and realize when they are considered separately. Thus, significance, classification and diversity of rhythmic motions, observed in the world ocean and atmosphere, may be indication for other to state that the solar cycles are also provided with the same features and directed to establish system's stability. In the same manner taking account of the fact that dissipation (scattering) and biological and geological rhythms takes place at every conservation of substance (energy), it may be suggested that such a phenomenon also takes place in hydrodynamic systems independent on internal scale of turbulence. Then, the similarity of cell structures of the ocean and atmosphere, the Earth's crust and the solar system (spatial periodicity) probably gives rise to the common character and similarity of their functional meaning. Finally, a similarity of cycle origin mechanisms may be an extra source of information in problems of formation and conservation of the solar system's cycles.

In this connection it should be noted that in the last decades problems of frequencies decay (splitting) and, especially, bifurcation are permanently discussed in physics particularly, in the field of nonlinear oscillations. For different systems, possibilities for frequency doubling and the facts related to gradual variations of system's parameters are established and revealed by theoretical and experimental methods. General laws of bifurcation (Faigenbaum's law and others) which are intrinsic to all nonlinear systems independent on their physical nature are found. It is found that frequency bifurcation phenomenon (differentiation or labor division) is also intrinsic to living systems, which as it seams is an additional evidence on the structural unity of living and nonliving systems and on optimum of processes proceeding in them. Rhythm splitting has a great biological meaning and significance for development and evolution of the biosphere. Within the frames of target-oriented changes of medium they are a

1. For example, to elucidate the glacial cycles' (**600 million years**) responsibilities within the cosmic cycles (events), as well as theirs preceding and succeeding cycles.

necessary condition to provide adaptability of living systems. In other words, rhythm splitting together with known changes of their parameters is the necessary condition for providing a stability of living systems. And in all that cases when changes in rhythm parameters do not provide and satisfy to relief and to dissipation of external load a frequency bifurcation takes place. A biological meaning of bifurcation resides in the fact that mechanisms, which use and process external energy resources faster and more efficiently, are arisen in living organisms, thus promoting increase of adaptability of organism, generalization of its stability and existence. The fact of frequency splitting here is also not continuous but occurs only at certain coincidence of changes of parameters of external off-loading and rhythms. Undoubtedly, causes of biorhythm splitting and their recognition will clarify correct perception and realization of such phenomena in nonlinear systems, and these causes are not so superficial.

Let us note in conclusion that many problems of cycles (structures, properties, forming, etc.) are in embryonic state and newly forming. We only attempted to reveal similarities that form their bases hoping that this phenomenon will attract attention of researchers and interesting problems concerning cycles will be studied proceeding from these generalities.

CHAPTER I

CYCLES OF THE SOLAR SYSTEM

The cycles of the Solar system represent regular changes in rotational and orbital movements of planets, satellites and other celestial bodies, as well as of their parameters and phenomena related to these parameters. These movements are closely related to the mass, the form, to orbital and rotational moment of momentum, center of gravity and the changes of its other parameters in the course of the time.

In order to understand correctly the essence of these cycles, their formation, course and the elements of structure, some knowledge about the Solar system are needed. Beginning from times immemorial, humanity was interested in structure of the Solar system, regularities of its movements, periodicity of these movements sizes and masses of planets (satellites), distribution of their orbital and rotational velocities. It is not less important to substantiate the causal or natural advancement of Solar System, etc.

The development of technical means, improvements of scientific methods have supported the achievement of success in processing the above mentioned problems. The scientists revealed several regularities of Solar system through detailed observances and investigations; some of them are shown in table 1. Thus, it becomes clear that distances between planets are subject to a rule, which acquired the name of its investigator e.g. Titsius, Bode.

According to that rule the distances between planets are expressed in the following form:

$$0.3*n + 0.4, \qquad n=0,1,2,4,8... \tag{1}$$

Then in accordance with calculations it is obtained that, the velocities of rotations of planets around their axes are mainly expressed by the formula

$$A = K*M^{2/3}, \tag{2}$$

where **A** is the momentum of the amount of the planets' movement per unit of mass, **M** is the mass, **K** is a constant. Such simple relations are also observed between the orbits of planets, the inclination angles, sectorial velocities and several other characteristics. Besides, these regularities have become very vulnerable under the pressure of new data, they also contain little information about the evolution and past of the system. It is important because during the evolution the material composition of the bodies, their mechanical chemical abilities undergo slow, irreversible (due to what?) changes, which reversibly affects the regularities and nature of their. That is why the existing comparable relations of the medium angle velocities of the planets (the satellites) are of special interest and contain significant information about the formation and reasons of the development of planets. The orbital periods T_1 and T_2 of two planets are said to be comparable, if their relationship can be written in the form of

$$T_1 / T_2 = n_1 / n_2, \tag{3}$$

where n_1 and n_2 are small positive numbers. If we were to construct a table of the orbital and rotational periods of the planet (the satellite) movements, it can be noticed that many of them are comparable with each other. Thus, the construction matrix of the Solar system containing classical comparability of 2:5 for the periods of Jupiter and Saturn has the following form (after Molchanov)

Planets' basic parameters and relations (after Alphven, Arrhenius) Table 1

		10^{13} cm	([c])	° ([c])	′	10^{6} cm/sec ([e])	10^{8} sec ([e])	10^{19} cm²/sec ([e])	10^{46} gcm²/sec ([e])
		10^{13} cm	([c])	° ([c])	′	10^{5} cm/sec ([e])	10^{8} sec ([e])	10^{19} cm²/sec ([e])	10^{46} gcm²/sec ([c])
Grazing planet [a]		0.00696				437	0.0001	0.304	
Synchronous planet [b]		0.253				72.5	0.0219	1.83	
Mercury	☿	[c] 0.579	0.2056	7	00	47.9	0.0759	2.72	0.906
Venus	♀	[c] 1.08	0.0068	3	24	35.1	0.194	3.79	18.5
Earth	⊕	[c] 1.50	0.0168	0	00	29.8	0.317	4.47	26.7
Mars	♂	[c] 2.28	0.0933	1	51	24.2	0.595	5.48	3.52
Jupiter	♃	[d] 7.78	0.0483	1	19	13.1	3.75	10.2	19 400
Saturn	♄	[d] 14.3	0.0559	2	30	9.64	9.34	13.8	7 840
Uranus	⛢	[d] 28.7	0.0471	0	46	6.81	26.5	19.5	1 700
Neptune	♆	[d] 45.0	0.0085	1	47	5.44	52.1	24.5	2 500
Pluto	P	[d] 59.0	0.2494	17	10	4.75	78.1	27.1	17.9
									$\Sigma \approx 31\ 507$
									Solar spin momentum[c] 170.

		Sidereal spin period, τ	Inclination of equator to orbital plane, i_{eq}		Mass, M	Equatorial radius, R	Average density, ρ	Normalized moment of inertia, $\alpha_{\Xi}^2 = R_{\Xi}^2/R^2$	Escape velocity, $v_{es} = \left(\frac{2GM}{R}\right)^{1/2}$	Ratio of semimajor axes, $q_n = \frac{a_{n+1}}{a_n}$	Distributed density, $\rho_{dis} = \frac{M}{a^3}$
		10^{5} sec ([c])	° ([c])	′	10^{27} g	10^{9} cm	g/cm³	([c])	10^{5} cm/sec ([e])	([e])	10^{-13} g/cm³ ([e])
Grazing planet [a]											
Synchronous planet [b]											
Mercury	☿	76.0			[f] 0.333	[f] 0.243	[f] 5.46		4.27	1.87	17.2
Venus	♀	210	~180		[f,g] 4.87	[f] 0.605	[f] 5.23		10.3	1.39	38.7
Earth	⊕	0.862	23	27	[c] 5.97	[c] 0.638	[f] 5.52	0.3335	11.2	1.52	17.7
Mars	♂	0.886	23	59	[f] 0.642	[f] 0.340	[f] 3.92	0.389	5.01	(3.41)	0.539
Jupiter	♃	0.354	3	5	[d] 1899.	[d] 7.16	[d] 1.31	0.25	59.4	1.84	46.5
Saturn	♄	0.368	26	44	[d] 568.	[d] 6.00	[d] 0.70	0.22	35.5	2.01	1.95
Uranus	⛢	0.389	97	55	[d] 87.2	[d] 2.54	[d] 1.3	0.23	21.4	1.57	0.0369
Neptune	♆	0.540	28	48	[d] 102.	[d] 2.47	[d] 1.66	0.29	23.4	1.31	0.0112
Pluto	P	5.52			[h] 0.66	[d] 0.320	[d] 4.9		5.24		0.0000322

[a] A grazing planet orbits just above the surface of the central body.
[b] A synchronous planet orbits with an orbital period equal to the spin period of the central body.
[c] Allen, 1963.
[d] Newburn and Gulkis, 1973.
[e] Calculated from equation given.
[f] Lyttleton, 1969.
[g] Howard et al., 1974.
[h] Seidelmann et al., 1971.

$$
\begin{matrix}
1 & -1 & -1 & -1 & 0 & 0 & 0 & 0 & 0 \\
0 & 1 & 0 & -3 & 0 & -1 & 0 & 0 & 0 \\
0 & 0 & 1 & -2 & 1 & -1 & 1 & 0 & 0 \\
0 & 0 & 0 & 1 & -6 & 0 & -2 & 0 & 0 \\
0 & 0 & 0 & 0 & 2 & -5 & 0 & 0 & 0 \\
0 & 0 & 0 & 0 & 1 & 0 & -7 & 0 & 0 \\
0 & 0 & 0 & 0 & 0 & 0 & 1 & -2 & 0 \\
0 & 0 & 0 & 0 & 0 & 0 & 1 & 0 & 3 \\
0 & 0 & 0 & 0 & 0 & -1 & 0 & 5 & 1
\end{matrix}
\tag{4}
$$

There is another regularity of comparability as well:

$$
\begin{matrix}
1 & -1 & -2 & -1 & 0 & 0 & 0 & 0 & 0 \\
0 & 1 & -1 & -1 & -1 & 0 & -1 & 0 & 0 \\
0 & 0 & 1 & -2 & 1 & -1 & 1 & 0 & 0 \\
0 & 0 & 0 & 1 & -6 & 0 & -2 & 0 & 0 \\
0 & 0 & 0 & 0 & 1 & -2 & -1 & 0 & -1 \\
0 & 0 & 0 & 0 & 0 & 1 & -2 & -1 & -1 \\
0 & 0 & 0 & 0 & 0 & 0 & 1 & -2 & 0 \\
0 & 0 & 0 & 0 & 0 & 0 & 1 & 0 & -3 \\
0 & 0 & 0 & 0 & 0 & 0 & 0 & 1 & -1
\end{matrix}
\tag{5}
$$

Theoretically, the vectors of frequencies, which are originated by these two matrices, are approximating the real vector with the same success. But it is difficult to conceptualize the physical causes, which bring to these matrices of frequencies. In celestial mechanics, the relations of the form of $\sum \mathbf{W}_i\ \mathbf{n}_i = \mathbf{0}$ between the rotational movements of bodies, are called synchronization or phenomenon of resonance, whereas in other sections of physics there is a difference between these meanings. The relations are the periods between of the rotational and orbital movements of the planets of the Solar system. The periods of the orbital movements of their satellites are resonance too. Considering two types of resonance are distinguished in the Solar system:

a). orbit - to - orbit resonance: When one orbit influences one or several other orbits, e.g. the planets of Mercury, Venus, Earth and Mars are in by the relation of $\mathbf{W}_M - \mathbf{W}_V - 2\mathbf{W}_E - \mathbf{W}_M = \mathbf{0}$, or the periods of Jupiter's and Saturn's orbital movements relate to each other as 2 : 5,

b). rotation - to - orbital resonance. This case is divided into two parts. This resonance that is caused when the rotational movements of planets influence their orbital velocities are called first order resonance. Thus, Mercury's rotation round its axis is equal to **58.8** days, which constitutes the 2/ 3rd of the duration of recycling. Resonance is called of second order, when the rotational movements of one planet affect the recycling of another planet (or satellite). Example, the rotation of the Earth influences the recycling of Venus and Moon. When the period of recycling of a body becomes equal to the period of the rotational movement, often this is called a synchronous movement. In the case of a resonance, relative positions of the bodies are connected to the particular states of equilibrium, round which they perform regular fluctuations. The nature of the movements in the case of a non-resonance despite the stability is much more complex. It is necessary to mention that in all cases only the average values of recycling, rotating, distances, interactions and other quantities are considered. The time and space averaging despite being efficient means for the mathematical investigations do not allow estimate the individual development processes of the celestial bodies, in this case.

These processes that originate from the transformation momentum of the celestial bodies and last until their destruction, represent regular, interrelated (by a straight and inverse correlation and regression) morphological, astronomical, power complex of transformations and transitions that are characterized by a particular time sequence. Thus due too such averaging the irreversible changes of the bodies are neglected in the course of time from the geological point of view. Still, the investigation of the resonating structure of a non-excited (without an external influence) system clarifies its evolution. Thus, the comparability of the average movements in the Solar system makes the causal nature of planets' and their satellites' average movements' distributions impossible. The transformation and preservation of the resonating relations mean, that environmental factors (conditions) and their changes have been repeated "identical" in the course of time from the geological point of view. How and where from has the synchronization of celestial bodies originated? There are two points of view concerning this problem. According to the first one, the bodies with their orbital and rotational movements and periods were originated without resonance. The last one was stated later, when the tides influencing these periods in different manners have caused resonating phenomena. According to the second one, celestial bodies have been originated under the influence of resonance. Otherwise, in the beginning only relations between bodies existed, which created resonance, and later due to the parceling of frequencies, new bodies and movements have originated. As a result of the increase of an external load, the mentioned way of origination of new resonance bodies, movements resonance (cycles) has many general features with the regulations of transformations, keeping, converting of biological rhythms (structures) which will be discussed in the appropriate section.

Usually, the planets' orbits are defined by six parameters, **a** - is the big semi-axis of circumference, **e** - is the eccentricity of orbit, **y** - is the obliquity of orbit to planet's equatorial plane, **Ω** - is the length of rising knot, **ω**- is the length of orbit's percent, and **ε** - is the century's mean duration. These parameters are constant if the movements are not excited and vice versa, variable when the movements are excited. Otherwise, if constantly influencing external forces exist then the parameters carry changes of different signs and values. The values that change in signs are repeated with some periodicity. Thus, the eccentricity of the Earth fluently changes in the range of $\mathbf{0.0007 < e < 0.0658}$ once in **89,000 - 90,000** years, receiving its greatest (smallest) value, the obliquity angle of the orbit alters in the range of $\mathbf{22^0\ 1 < \varepsilon < 24^0\ 5}$ and once in each **25000= (23,000- 28,000)1/2** year the greatest (smallest) value is obtained, the wobbling of equinox alters in the range of $\mathbf{42^0 < n < 48^0}$ and each **40,000=(35,000 -45,000)1/2** the values of it are repeated. Such intentions are demonstrated by the changes of big semi-axis of the orbit, the length of rising knot and other parameters. This regularity is also preserved in the range of the rotational parameters. Celestial mechanics yet is not able to give a correct sufficient answer to the reasons of changes, nature, possible relations of that parameters. The equations of the changes of the above-mentioned parameters have great significance in the definition and specification of the orbits of celestial bodies. It is not randomly that they are very basic in the celestial mechanics. The great scientists of the past, (such as Newton, Lagrange, and others) were engaged in spreading of these equations and they received the basic results these problems by different ways. In the investigations always it is considered that the external excitements are functions of time and the same parameters **R=R (t, a, e, y, Ω, ω, ε).** Their form is different depending upon the conditions applied to the components of force and parameters. The following form of the system is one of the particular cases:

$$\frac{da}{dt}=\frac{2}{na}\frac{\partial R}{\partial t},\qquad \frac{de}{dt}=\frac{1}{na^2e}\frac{\partial R}{\partial a},\qquad \frac{dy}{dt}=\frac{1}{na^2y}\frac{\partial R}{\partial \Omega} \qquad (5)$$

$$\frac{d\varepsilon}{dt}=\frac{2}{na}\frac{\partial R}{\partial a},\qquad \frac{d\omega}{dt}=\frac{1}{na^2e}\frac{\partial R}{\partial e},\qquad \frac{d\Omega}{dt}=-\frac{1}{na^2e}\frac{\partial R}{\partial y}$$

where $\lambda = \int_0^t \mathbf{n} dt$, **n** -is the average movement of the satellite. It is known that, analytical solutions (the inte gral) of these equations are impossible to obtain. That is why, their approximate solutions use series, as is usually done in such cases. But the above mentioned solutions do not clearly indicate the solutions of **a, e, ω, y,** parameters etc. it is difficult to separate those. Probably the reason is that the exciting of function **R** (or the time) effects all parameters identically, meanwhile in reality that is not true. Then the excitement being external must be independent from the system. It is necessary to define not all components of the excitement theoretically possible, but only those that generate and originate constantly particular periodic changes. Otherwise, the investigations of hierarchical abilities of the regular changes of different scale will be changed into the analysis and conjugation of series. Such approach substitutes the natural problem to a pure formal problem. Related to this an interesting idea the famous evolutionist has been mentioned that the difference between a theorist, an experimentation and researcher working in the field is the following. The theorist considers all possible hypothesis without an exception, the experimentation considers only the allowed hypothesis and the researcher is limited by the real hypothesis. A theorist must ask the experimentation whether his/her hypothesis is of some interest and the latter should ask a researcher whether it has something to do with the real hypothesis. The following relation relates the celestial bodies' revolutionary angular momentum to parameters of their orbits:

$$\mathbf{c} = [\, \mathbf{a}(\, \mathbf{I} - \mathbf{e}^2)\,]^{1/2} \qquad (6)$$

Consequently, when **a, e, ω** change under the influence of the external appropriate forces, they create a change of the angular momentum of recycling of the bodies. If there exist any change in the angular momentum, then it should be transferred either to another body or to the interplanetary environment (which, as it is known, has very little density and consequently, the transition in that direction will be insignificant). A part of the amount of the movement is transferred to another body through the orbit - to - orbit resonance (if such a relation exists), another part is transferred to the rotational momentum through the rotational - to - orbital resonance. This brings to changes in the rotational angular velocity momentum of inertia with respect to the rotational axis. This, latter, in turn, cause plastic or elastic deformations in the surface layers of the body, of the distributions of heat, stress, density and movements inside it. Thus, a gradual redistribution and dispersion of the energy of the initial excitement takes place during this period. This hypothesis is not new, but it is not complete either. Let us try to substantiate it. Thus the continuous transition of the momentum between the Earth and the Moon doesn't cause any suspect. That problem is well processed and by some investigators, the Moon was a separate planet in past, which later was captured by the Earth and became subject to the movements of the Earth. Thus, the Moon is a captured planet, which has been brought to its present orbit, and placed there by the influences of tides. Then, according to some sources, the transitions of the amount of movement have taken place everywhere in the Solar system if not now, then at least during of some period of it's evolution. The basis of that is the same direction of the rotational and orbital movements of the majority of the central (Sun's), the secondary (planets'), and the tertiary (satellites) bodies.

According to the Alphven, the satellites moving in the backward directions are captured by other orbits (according to what - according to complementary and uncompensated momentum of the amount of movement). The Venus backward rotation is considered was formed in the period of the origination (matter accretion) of the planet, but that explanation is not final. According to that sources, the changes of the Sun's momentum of the amount of the rotational movement are transferred to the recycling momentum of the amount of movement, the changes of which are transferred to the recycling momentum of the amount of movement of the planets, the changes of which cause changes of **a, e, w, ω** parameters of the orbit, as well as changes in the rotational momentum of the amount of movement. In its turn, the originated additional energy generates deformations in a point of body, redistribution of thermodynamically quantities, some part of which is transferred to the recycling momentum of the amount of movement of the satellites. As a response to the transferred momentum of the amount of movement, the orbital parameters of the satellite carry out appropriate changes with the help of which the part of energy received, is dissipated and prevented from the effects. The movement's amount, which is not liquidated, is transferred to the rotational momentum of the amount of movement. Thus, during the period changes of the orbital momentum of the amount of movement are consequently distributed to its momentum of the rotational movement, the orbital and rotational movements of the secondary and then, tertiary bodies decrease and dissipated gradually. This hypothesis is also substantiated by the principle of hierarchy of the structures. As it is known, in such cases the objects' activities is spread and took place under the pressure of high level interactions. They, in their turn, become an environment, where the tertiary bodies operate, etc. The hierarchy is not only a time - to - space one, but also a functional one, i. e., the role and meaning of the structures in the whole volume of the work, is definitely subordinated. The allocation of the central, secondary, tertiary and other bodies are different and conditioned by the galactic (solar) division of the work. What facts do exist for such conclusions? It is seen from Table 1,that planets' rotational momentum of the amount of movement constitutes only an insignificant part of their orbital momentum of the amount of movement. The same is noticed in the momentum of the amount of rotating of the satellites and the rotational movements of the planets. The sum of the momentum of the amount of orbital movements of the planets and the momentum of the amount of rotational movement of the Sun are approximately the same. The energy part, which is consumed on the activity of the Sun (regular movements and redistribution's inside it and the surface layers) is not included here. Then, according to the investigations, the planets have transferred to the satellites only a little part of their momentum of the amount of rotational movement in case when the Sun has transferred to the planets a great part of it's momentum of the amount of rotational movement. Otherwise, the planets dissipate the movement's transferred amount basically and only an insignificant part of it is delivered to the satellites. Hence that explains their small sizes. If there were be an addition (shortage) in momentum of the amount of movement, then it would bring to a new condensation of the interplanetary substance in the course of time (a transition to another orbit). A transition of the momentum of the amount of rotational movement of planets to their satellites' changes the axial (orbital) rotations in the range of a few percents of their value. The resonance relations of the present rotational and orbital movements have not been formed invisibly, but have had their predecessors. They are stipulated by the universal (galactic) environmental conditions, which according to the observations, are not stationary and, even more, have some trend. Consequently, for the Solar system to preserve its state of equivalence in the variable environment, it is necessary to rebuild the previous resonating relations (fluctuations) in order to prevent from the effects of the new conditions in the best way. The fact that the resonance relations and inter-influences of the Solar system have had their predecessors (which significantly had differed from the present relations) is testified by the acceleration's, in particular, tectonic cycles' phenomenon, which clearly appears on the geological axis of time during the investigation of tectonic processes.

The nature of tectonic cycles, their structural time - to - space specifications will be discussed later. Here, let us only mention, that according to the existing geological data, the duration of tectonic processes demonstrate some tendency in the course of the geological development of the Earth, that during the time of their periods are gradually shortened. It follows from here that the causing reasons (forces) of that also must demonstrate such tendency or ability. This means that, the velocities of the Earth's rotating and recycling, to which the tectonic processes are related, went on comparatively longer and slower in the deep past. Consequently, the quantities of resonating relations and inter-influences in the Solar system had another distribution on the geological axis of time. In particular, their resonance layer was wider which can be a basis for the Molchanov's hypothesis.

In the theory of the non-linear oscillations exists the idea of the boundary cycle to which all the orbits of the system gradually urge (or vice versa, from which gradually move away) in the course of time. A conclusion should be made that the geological cyclic changes being the directions of the movement of the appropriate dynamic system approach their boundary value in the course of time, after which they will not change. And so, the misgiving of some geologists that the acceptance of the acceleration can bring to a deadlock is groundless. According to Darwin the main reason of the changes of rotations is the influence of tides, which during a long period of time can change basically **a, e, ω**, etc., orbital element of the heavenly bodies. But by Alphven and others the influences of tides on the orbits of the planets are very little to be significant for the evolution. The transition of the planet's moment of momentum to the orbital movement of the satellite is performed through the tangential component of the potential of the tide. But tide's radial component, despite to not creating a momentum of the force, realizes the dissipation of the mechanical energy of the system. Consequently, decreasing the orbital energy causes a decrease in the eccentricity of the satellites by itself. And in general, the entire changes of the eccentricity of the satellites, which are stipulated by the momentum of the tides and radial forces, are positive if the rotation of the planet is much faster than the recycling of the satellites. Exactly those define the direction of the evolution of the tide of the orbits of the satellites. Thus, the tides can influence on big semi-axis of the orbit of the satellites, on **a, e, ω** other parameters of the orbit of the planet, on its obliquity of the rotational axis. But which one of those is primary, and what principle of subordination does exist between them, is not clarified finally. From the mentioned data a conclusion can be made that the significance of the Solar system and its members is not limited by the Sun's and planets' (as well as the planets' and satellites') equivalence or balance of the momentum of the amount of movements., when considered as a whole, perform a function that yet is difficult to formulate completely. One thing is clear, that its significance is stipulated by the galactic labor's division, which always carries changes in the course of the time. What does go on with the waves of a tide and inter-influences? At first, the specialists considered that those are wholly dissipated during the cycle by the permanent movements of the shallow and beach waters. But later the calculations showed that the changes do not provide the braking of the tide on the Earth. Other possible processes of the tide, that are originated inside the Earth, in the massive of land and ocean, and cause geological, ocean, atmospheric cycles, have their inculcation in the overall loss of energy.

Indeed, when the body rotates in a field that causes tide periodically, it is deformed relatively to the rotating coordinate system by a period of 1/2. The deformation creates stresses (elastic, thermal, and etc.) redistribution of the density, temperature, material transitions, (tectogenesis) which under the external periodic influence (establishing the stability of the system) becomes cyclic and represent dissipaters of the energy of deformation.

The problem of the dissipation of energy of the tide in the massive of the land, water, circulation of the atmosphere, which is not finally clarified and estimated, is of a special interest. In order to reveal the possible influences and periods of the waves of the tide, the scientists transfer its force They potential into the harmonic series. First, this problem was entirely solved by Dudson in 1921. The series proposed by him include three basic groups of periods and 385 strictly harmonic members. The long - term members of the Moon's and Sun's force of the tide are presented by 99 independent members of the series the whole daily periodical members constitute 157 members of series, and the members of a half daily and small periods constitute 129 representatives of the series. The changes of the force of tide have a complex nature in the sense that the distribution of the energy to the mentioned members is not identical. Thus, if the members of long periods are grouped in the following way, of 7,9, 15,30 days long, a year and multi-century duration, then the distribution of the potential force the mentioned components by a percentage of it to the total is:

1. long-term members **10.71%**
2. members with one and a half year period **14.24%**
3. members with one month period **22.31%**
4. members with a half month period **41.11%**
5. members with nine days period **9.38%**
6. members with seven days period **2.25%**

To the inter-influence time scales correspond space dimensions (formation, structures), through which the fluctuations of the appropriate period are realized. The separation of these space scales is done on the basis of the characteristics of the Moon, Sun, Earth, as well as on the characteristics of their circumferences' mutual positions. Not to make the description boring let's consider only a few representatives of those:

$$W_0 = +\,0{,}50458.0{,}5(1-3\sin^2\varphi)\cos\theta'$$

$$W_1 = -\,857{,}0(1-3\sin^2\varphi)\cos N' \qquad \text{nineteen year old}$$

$$W_2 = +\,153{,}8(1-3\sin^2\varphi)\cos(h-p_1) \qquad \text{one year old}$$

$$W_3 = +\,953{,}1(1-3\sin^2\varphi)\cos 2h \qquad \text{half a year old}$$

$$W_4 = +\,206{,}4(1-3\sin^2\varphi)\cos(2h+p_1+S) \qquad \text{one month old} \tag{7}$$

$$W_5 = +\,179{,}2(1-3\sin^2\varphi)\cos(2S-2h) \qquad \text{half a month old}$$

$$W_6 = +\,74{,}4(1-3\sin^2\varphi)\cos(3S-2h+p_1) \qquad \text{nine days old}$$

$$W_7 = +\,62{,}5(1-3\sin^2\varphi)\cos(4S-2h) \qquad \text{seven days old}$$

Here $N' = -N$ means the width of the originating knot of the Moon, S is its average width, **h** is the average width of the Sun, p_1 - is the width of the perigee of the Sun. The first member of the series W_0 is a special one, it represents permanently acting forces of the deformation parallel to the periodic force of the tide on the surface of the Earth. Here **G** is a gravity coefficient

$$G = \frac{3M\rho\, a^2 r^2}{4E\, c^3} \tag{8}$$

For the Moon it is equal to $G_m = 26{,}16\ cm^2/s^2$, and for the Sun $G_s = 12{,}05\ cm^2/s^2$. Having the largest period of the fluctuation and the largest space scale of it is the simplest representative of the potential of the gravity force. The mentioned scales of the space, except the influence of the appropriate component of the tide's force, evidently fluctuate under the influence of other forces (in particular of the earthly origin) and support their dissipation. Thus it is seen from the above mentioned that the potential of the tide represents a summary of influences originated from different confrontations of the Moon's, the Sun's, and other, near the Earth bodies' mutual positions of the orbital and rotational movements. Consequently, the same external potential force (depending on distribution of its period) can have its contribution in the origination of the differently scaled cyclic changes. Thus, for instance, it can take part in the earthly daily tides and at the same time in the periodic changes of wobbling and nutation of the Earth's rotation axis in the space. And the opposite, each cyclic change can be simultaneously under the differently scaled influences. That is why the transformation of the cycles of the Solar system, the strict separation of the forces and reasons of its movement is related to big difficulties and complexities.

Thus, in the Solar system the influences of the tide play a significant role in the rotational and recycling changes of the satellites of the planets and other bodies. Such changes can take place through a resonance (

but in that case only a little part of the momentum of the amount of movement is transferred, much enough to ensure the resonance relation between the bodies), as well as when the gravity potential of planet will include multipliers of the type **$(1\text{-}3\sin^2\varphi)$**, i.e. will be dependent on the geographical width. It means that, the gravity potential must have a tangential component, which $\neq 0$, and it is necessary for the planet to be a little flat for that.

It is seen from the short description of the cyclic changes in the Solar system that they don't have constant periods, but move within some (allowable) range. The movements of this kind seem causal, out of any regularity and do not attract the attention of the specialists. Rather, the average values and quantities or periods are defined and evaluated, which are successfully utilized and satisfy the practical requirements. In the same way the average velocities of the rotational and orbital movements of the planets (satellites), the six elements of their elliptic orbit are calculated and specified. But it is necessary to mention that the changes of the periods, no matter how they seem insignificant, are originated arbitrarily, but each time (i.e. during each cycle) their sizes and signs are defined and regulated by the external forces, by their "dictation" and "requirement". Due to such movements, the response to the external, and necessarily also periodical, change is the stable state of the system. That is why the possible movements of not essential and insignificant, from the first point of view, parameters of the cycles in the Solar system have a fundamental and immense significance in the evolution of the Solar system. In particular, the adaptive movements of a satellite (planet) relatively to a planet (sun) are constructed and provided by that movement. Moreover, the adaptive "abilities" of a planet (satellite) are defined just by the largest value of that variation, their embrace and depend on them. It should be accepted that the deviations of the parameters of the solar cycles have been defined or, more precisely, have been planned in a form of an order for a proper amount of the movement to be transferred at the proper time and in the proper place. And the system (a planet, a satellite, an asteroid) should be ready to anything that is originated in the external and also planned environment in order for a valuable change in the Solar system not be imperceptible and be considered for the preservation of its stability. It is evident that if the parameters of the cycles are fixed then the acceptance of the amount of the movement, its further transition and dissipation between the celestial bodies during a cycle would not be possible, which naturally would cause a non stability. Thus, the deviations of the parameters of a cycle are of an exceptional significance for the stability of the Solar system and specially, for the investigation of its evolution. Their discussion probably will allow to understand the alternation of the forms of the stability of the Solar system (planets, satellites, and etc.) depending on the conditions of the galactic environment. Further, there is a spatial subordination in the already discussed cycles. The orbital and rotational regular changes of the tertiary bodies take place in the environment of the appropriate changes of the secondary bodies (under the conditions). The secondary bodies, in their turn, go on in the environment of the same type of movements of the central body. The orbital and rotational movements of the latter and their changes take place under the conditions of the galactic environment and so on. Usually, such a picture of subordination is seen also when the periods of the time-cycles include one another. But here such a intention is not formulated clearly. Thus, if the orbital movements of the Sun, Moon, Earth demonstrate a subordination in the time like **29, 365,6.10^{11}** days, but this is absent in the **28** days, **24** hours, **25** hours of their rotational movements.

What are the explanations to this? Let us bring some of them:

a) the variety of the conditions for the formation of the cycles,

b) the violation of the transition order,

c) the rareness of the state of the Moon, and etc.

Let us mention one more important thing. In the structures of the complex hierarchical systems (Solar system is one of those) the important thing is not the synchronization of the movements, or the existence of a resonance of its separate links (bodies), but the concordance of their activities in the course of time or, as it is common to say, the coordination. In such cases, the constant relations of the periods or frequencies are not obligatory. Some separate links can fluctuate synchronously or as- synchronously, be in a resonance or not, here the basic thing is the satisfaction of the optimality of the activity of the system. In any case the energy of the excitement must be distributed in an optimum way between the links. The establishment of this or that mode of the activity between the oscillators depends on the structure of the external excitement. If the excitements of the environment or their sequence are constant in the course of time, then it is possible for a resonance to be established between the movements and activity of the system to satisfy the criteria of the optimality. But the excitements and their sequences are changed, consequently, the condition of the stability will require a coordination of the activities of the links. It is clear that the coordinated movements will preserve their optimal significance until the conditions of the environment have not changed. After the change, the conditions will lose their significance, and the coordinated movements appropriate to the new conditions will be formed. Usually, the coordination is realized for some special purpose or in the connection with a problem. The activities of the links of a system are coordinated for the general problem, in a way that they wholly will realize the purpose or the problem of the system.

CHAPTER II

THE SUN CYCLES

A study of Sun cycles is very important, in the first place, to understand correctly the physics of the Sun, its structure and inner dynamics. Sun cycles are also important in order to and evaluate the laws of development of and the motions of the solar system, their effects particularly on the Earth and especially on the biosphere. The spectrum of observed Sun cycles is investigated extremely in homogeneously. Its high frequency oscillations are investigated in great detail. The specific features of the methods applied cause it; in order to record and study the Sun cycles, telescopes with observation time intervals not exceeding one day are used. Let us subdivide conditionally the Sun cycles into three groups. The first group is the cycles with regular changes lasting from fractions of a second up to one year, the second group is the cycles lasting up to 100 years, and the third group is the cycles of other frequencies. The detection and study of low and intermediate frequency group cycles are difficult because it requires long observation intervals, which as it was noted are difficult to achieve.

The existence of such cycles on the Sun is certain. The methods for their detection and study for the most part differ from the classic ones. Investigations of these cycles are indirect and thus one must study the traces left by these cycles. Particularly, these cycles' corresponding effects on various phenomena and formations (climatic, air and water masses, sediments, geologic life, biosphere). In the globe the roles and significance of the cycles of three types observed in the Sun's activity and motions are different. The intermediate and low-frequency cycles are of interest to us since they are related in one or another form to the structural formations of some surfaces (photosphere, ionosphere, etc.) of certain spatial dimensions and different scales. The latter give opportunity to consider the problems of structure and classification of the Sun cycles and, hence, partly of stability and evolution of the system in the domain of non-stationary excitations. The Sun's intermediate and of low-frequency c ·cles, according to data in the literature, are studied to different depths. Some of them are given in Table 2. where some of the characteristics of these cycles are indicated.

Of the Sun cycles the most known and studied are 11-year cycles. From the view of point of the solar physics the cycles of other orders are of no less important. What are the Sun cycles taken separately?

An 11-year cycle of the Sun is the period of activity and represents change in the relative number of the sunspots of a particular group in the course of time. The German scientist Wolf was the first to count sunspots, and compare different series of magnitudes of sun spot numbers. He found that these series almost repeat every 11year. These series were named after Wolf in his honor. The series of Wolf's numbers are very convenient for investigations but at the same time these series have one big drawback. These numbers have no physical meaning and their determinations are accompanied by a number of subjective factors, such as individual approaches in separation of sun-spot groups, differences in observations and instruments counting of isolated nuclei, etc. Taking these facts into account, and in order to study the Sun activity, the scientists propose to use more reliable indices: summary area of sunspots' surfaces and mean life of sunspots. Taking account of the magnetic nature of the 11-year cycle of the Sun activity, it is considered more expedient to use changes of its magnetic field strength. According to the recent investigations this approach is also not reliable enough. Since the 11-year solar activity cycle influences are on the long-term variations of mean level of the ocean, the ocean icing, the atmospheric cycle and a number of other phenomena of the globe certify. Thus the preceding facts show that the 11-year cycle's origin is not only magnetic in nature, but is more complicated.

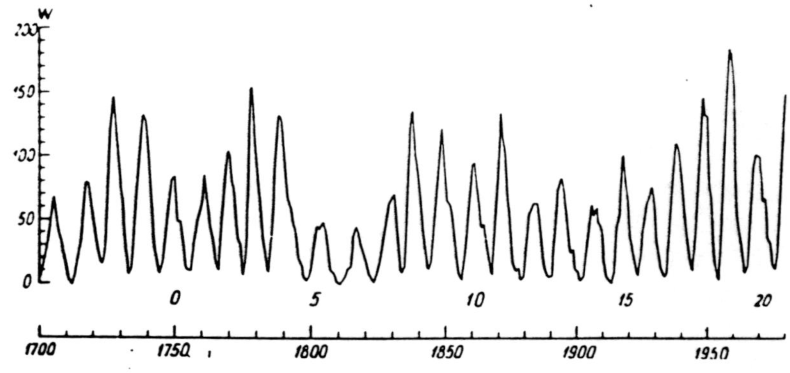

Fig.1. Average annual changes curve of Wolf's numbers for 1700-1979

What specific features do the Sun 11-year activity cycles have? According to investigations they do not coincide when imposed one on another and each cycle is represented as an individual process. Within their periods no symmetry is observed, and even more, within each cycle two completely different physical processes are distinguished. The former represents rhythmic outbreaks of activity, and the latter, processes due to this outbreaks' disposal. The first ones characterize growth domains (stages) of activation nucleus period, and the second ones characterize decreasing domains (stages) of cycles. The first have "explosion" nature, and their causes are located in the photosphere, where they are formed below the lower layers, and the second ones represent processes of relaxation and restoration nature and are due to turbulent dispersal and compensation processes in the photosphere and its upper layers. There is an inverse proportional relation between the cycle's intensity and its duration, i.e. the higher is the highest value of the sunspots relative number, the lower is the cycle duration.

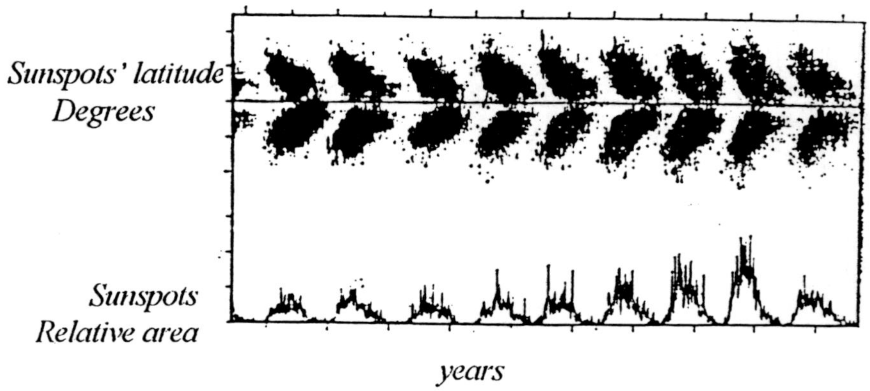

Fig.2. Arrangement of the sunspots in its corona and surface occupied by them vs. time

Table 2.

Basic Cycles of Important Phenomena in Lithosphere, Hydrosphere and Atmosphere, Which is Related to the Sun Activity (after Eigenson)

Author and year	Approximate period	Type of phenomenon	Phenomena for which cycles of change are confirmed	Epoch
1	2	3	4	5
1. Klaf(1905-1925) 2.Shostakovich (1931 and 1934)	3 2.6 -3.3	Geophysical geophysical geophysical climatic	magnetic fields strength meteorological sunspots, elements of climate silt sediments	
3.Angengaister	3	meteorological	air pressure and temperature in Samoa	
4.Angengaister	2.7	geological	slit sedmints in lakes and oceans and geological sediments	
1.Kalf (1920)	7	solar activity	climatic changes related to solar activity	
2.Shostakovich (1931)	5.7	heliophysical	the same as for 3 year cycles	
3.Shostakovich	5.7	geological	the same as for 3 year cycles	
4.Lungers hansen		geological	strip-like clay sediments of	during 3 tc 4 mln

			Southern Ural	years or 500 years
1.Schkabe-Wolf (1861)	11	solar activity	sunspots	
2.Peterson (1914)	9-13	solar activity	sunspots	
3.Helman (1906)	11.4	meteorological	precipitations	
4.Bauler (1926)	11	heliophysical	changes of air temperature related to Wolf's cycles	
5.Shostakovich (1931)	10.4 - 11.5	heliophysical	the same as for 3 year cycles	B.C
6. Perfiliev (1926)	10 - 11	geological	slit sediments of Like Saksi	
7. Shvedov (1941)	9	biological	strength of trees	
8. Douglas	11	biological	strength of trees	before our days
1.Shostakovich (1931)	16 - 18	solar and geophysical	sunspots, magnetic fields strengths, meteorological and hydrological	
2. Wagner (1924)	16	climatic	precipitations in Rome	
3. Le Dauna	17 - 18	hydrophysical	temperature of	

			oceanic water	
4. Le Dauna	18.5	hydrophysical	oceanic tides	
1. Hel (1919)	22	Sun physical	sunspots	
2. Johnson (1946)	20	Sun and geophysical	precipitations related to sunspots	
3. Gillet (1938)	20	geological	strip clay sediments	
4. Douglas (1938)	21.2	biological	intensity of tree growth	
5. Schulman (1942)	21 - 24	biological	intensity of tree growth	before our days
1. Fritz and Lokiet (1893 - 1901)	35	solar activity	sunspots , number of severe winters , etc.	
2. Brikner (1890)	30 - 40 cli matic and hydrologic cal precipitati ons, air temperatu re , variations in lake levels			
3. Brikner (1904)	30 - 40	climatic	Icing conditions of Alps	
4. Gann	35	meteorological	precipitations	

(1902)				
5. Shostakovich	28 - 31	Sun and geophysical	sunspots, meteo-logical elements, level of water basin	
6. Shostakovich (1949)	27 - 47	hydrological	regimes of isolated lakes	Past 500 mln years
7. Lunges hadusen (1946)	30 - 35	geological	strip-like clay sediments of Southern Ural	
8. Douglas (1909 - 1919)	29 - 33	biological	tree growth strength	1413-1906
1. Gouncki (1904)	73	Solar activity	sunspots	
2. Peterson (1914)	90 - 92	Sun and geophysical	sunspots and meteorological phenomena	
3. Memery (1935)	100	Sun and geophysical	sunspots and meteorological phenomena	
4. Glaisbery, Aigenson (1942-1946)	80 - 90	Solar activity	sunspots	
5. Kepen (1918)	89	Meteorological	air temperature of Wester Europe	
6. Kepen	93	Hydrological	oceanic tides	
7. Perfelev (1926)_	70	Geological	slit sediment of Like Saksi	
8. Gillet	73	Geological	strip-like clay	

(1938)			sediments and others	
9 Douglas (1931)	101	biological	tree growth strength	
1. Pretdechenski (1944)	189.4	solar activity	Sunspots	
2. Brikner (1904)	160	climatic	icing conditions of Alps	Glaciatio n epoch
3. Gillet (1938)	189	geological	strip-like clay sediments	
1. Fritz (1978)	300	Sun and geophysical	frequency of northern lights	
2. Klaf (1905)	300	solar activity	duration of Wolf's cycles	
3. Shnitnickov (1949)	200-400	geological	highest variation of lake slit sediments	B.C. A.D.
4. Douglas (1909)	280	biological	tree growth strength	

Then, the cycle's increase stage is shorter by 2 to 3 years than the duration of its decrease. For weak cycles the domains of growth and decay are equal, and for super weak cycles the opposite phenomenon takes place, i.e. the growth stage is longer than the decay stage. For this reason 11-year cycles do not represent a superposition of any harmonic components. There is a particular and sufficiently strict link between them, which does not resemble links of expansion series. Physical independence of individual cycles does not to allow us to consider them as statistical featureless elements of any set. The Sun 11-year cycles (and not only of that order) are not provided with strict physical periods. Their periods vary within the limits from 7 to 14 years. Each 11-year cycle represents a more or less sudden pulse magnification of the Sun activity with respect to some average state. At the end of the 11-year pulse the Sun is at its initial state again. The distribution of sunspots over its surface is not uniform, the changes of its groups with respect to latitudes obey the so called Shperera's law and play an important role in the problem of activity changes' evaluations. According to some authors, displays of the Sun 11-year cycles in different hemispheres (the northern and southern ones) are the same, and according to other sources they have some divergences. The cycles in the equatorial zone and in higher latitude zones (**±30□ to ± 60□**) are not the same. All these facts evidence about multi-cycle nature of 11-year cycles of the Sun activity. It is also indicated by 11-year oscillations of:

a) the summary power of the Sun activity in both hemispheres;
b) the relative summary activity in each hemisphere;
c) the average power of the Sun activity;
d) the geographical distribution of active latitudes;
e) the geographical distribution of active meridians; and
f) activity bursts and other phenomena.

If 11-year activity cycles are arranged study in chronological order it may be found that the magnetic field of primary spots for odd numbered cycles in the northern hemisphere is positive (northern direction), and that for even numbered ones is negative (southern direction).

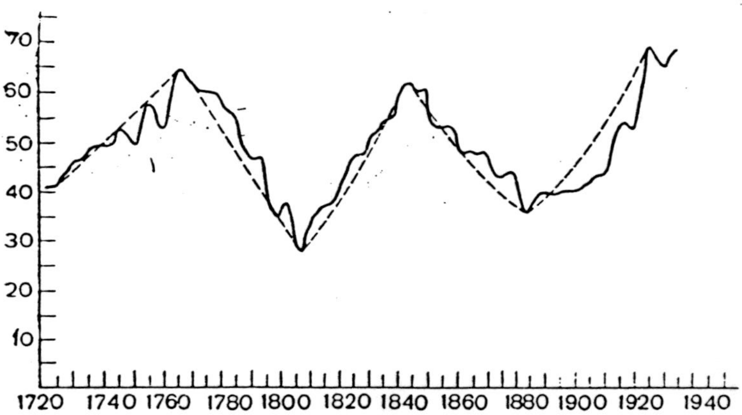

Fig.3. Wolf numbers 80-90 year cycle

The opposite pattern is observed in the southern hemisphere. Thus, the change of magnetic field polarity on the Sun takes place once in 22 years. This change forms the basis of 22-year cycles of Sun activity. Summary and maximum magnitudes of sunspot number variation possess the same periodicity of changes. According to some sources, changes of magnitude of sunspot motions along the Sun's diameter and along its meridian also have 22-year periods. The structure of these cycles is the same as the structure of 11-year cycles. The higher is the power of 22-year cycle the shorter is its duration. The duration and power of 22-year cycles are regularly changing in the course of time, periods being from 80 to 90 years. The series of the highest magnitudes of 11-year cycles, or the series of mean annual sums of the Wolf's numbers is also changing with approximately the same period. But it is not completely understood which physical phenomena cause such changes. 11-year cycles' are involved in century cycles and these cycles are one of century cycle's elements. Taking this fact into account it is suggested that the duration of century cycle is not constant but is changing within 65 to 105 years. The following series of the Sun cycles (in years) is often mentioned in the literature:

0.78, 1.05, 1.09, 1.09 -1.21, 1.3 -1.5, 1.47, 1.8 -2.4, 1.96, 2.1, 2.2, 2.3, 2.5, 2.8, 3.1, 3.4, 3.5, 3.53, 3.6, 3.9,4.1, 4.47, 4.6, 4.79, 5.0, 5.5, 5.5-7, 5.6, 6.0, 7.9, 8.3, 8.9, 9.3, 9.6, 9.9, 10.3, 10.45,

10.47, 10.5, 10.6, 10.7,10.79, 11.1, 11.13, 11.8, 11.9, 12.9, 14.1, 15.9, 16.0, 16.2, 17.6, 17.75, 17.8, 20.3, 22.0, 22.11, 22.4 26.2, 30, 35, 44, 55, 57.8, 65, 75, 79, 80, 91.3, 169, 176, 178, 189, 350, 400, 600, 700, 900, 1000, 1700-1900.

Regular changes of geological, climatic, hydrological and other natures related to 520, 600, 1400, 1500, 1700,1800-year and other periods of the Sun activity, which were found by different specialists, exist too.

The multi-cycle nature lacking a physical basis results in the notion of an infinite number of Sun cycles, loss of the problem's meaning and value. Therefore, when distinguishing geophysical phenomenon of any duration as an indicator for the Sun cycle of corresponding order, it is necessary to get assured, whether cycle of such order exists in the Sun. The largest regular changes of the Sun surface layers according to scale and amplitude are the result of the Sun's stratified rotation. Moreover, they rotate faster in the equatorial and low latitude zones: the period being 25 days, than in higher latitude zone: the period being 29 days at the latitude of 60°. According to some assumptions, except for regular motions along latitudes, there are also meridian motions of matter on the Sun. Contrary to 11- 22-year and other magnetic cycles, they are of mechanical origin and are not studied completely.

Studying the nature of the Sun cycles and the similarity of their formation, specialists derive certain general regularities of their structures and behavior. Thus, a group of authors states that regular changes of different scales occurring on the Sun represent a close set of cycles related to each other. This set forms a large, huge system of physical processes and has a particular subordinated space-time structure, based on the following circumstances: a) century cycles occurring on the Sun comprise its 11- and 22-year cycles. The same considerations are applicable to 22- and 11-year cycles, which in their turn comprise 3-4 year cycles. Higher frequency cycles are formed and developed on the background of lower frequency cycles, being in a certain sense their continuation; b) the Sun activity cycles of all orders according to their morphology are qualitatively similar to each other. Details of their forming, development and behavior are similar to each other and uniform; c) duration of every next cycle is greater by a factor of 2 to 3 than the proceeding one; the same is found in the space scales. According to this view of point the causes of the Sun activity rhythms origin are uniform and moreover they have hereditary links of origin. In this sense the mechanism of their origin is singular. It is supposed that this mechanism is related to energy of the Sun and its motions and redistribution. Hence, the Sun activity cycle is a multi-cycle process. It means that the cycles of higher order are composed of the lower-order cycles, and the former are not passively imposed on the latter but actively affect them. It is very clearly exhibited in the sphere of 11-year cycles. Thus, they are attributed to a number of structural formations on the Sun surface, particularly, in the photosphere and the chromosphere above it. These formations are solar granules, sunspots, bursts, solar prominence, flakes, and so on. The Sun activity includes all the above mentioned formations. Among these formations, certain sub-ordination exists and -regular interactions are present.

Granules are small structural units and component elements of sunspots and other formations. The latter ones, in their turn, are the elements of larger formations. They have the following time and spatial scale distribution: granules are **350 - 500** km long, sometimes **1000** km, their life lasts 5 to 10 minutes (20 minutes); sunspots are **100-4000** km long (with life lasting from 1-2 day to 12 month); prominence are **500 to 15000** km long (they exist approximately 20 days), size of flakes depends on the observation spectral line (flakes exist from 200 to 900 days). Bursts last from 2-3 minutes to 2-3 hours. A total number of sunspots and prominence located in the photosphere, as well as flakes located in the chromosphere, is changed depending on the stage of activity cycle, and regular transitions from prominence to flakes exist. But nevertheless inter- relations between the mentioned formations are not determined sufficiently accurately.

The regularities of their distribution are interesting. The most reliably determined are the distribution of sunspots. It is found that they arise not in the whole surface of the Sun, but on both sides of the equator at the latitudes $\pm 35^0$ and that their number gradually grows with the cycle development and then decreases until equator they occupy latitudes ±8°. Distributions of granules, prominence and other structural elements are not clarified yet. Thus, it may be said, with some restrictions that trends of particular subordination relations and transition directions are really outlined among space-time scales of the Sun activity cycles. The viewpoint of the other group of researchers that the cycles occurring in the Sun are included into each other and have a common cause is not correct. According to them, cycles and activity perturbations cannot be considered as rhythms. They propose five classes of oscillations: random oscillations, excitations, rhythms, periodic oscillations and periodic excitations, but the difference between oscillations of these classes and which one is classified as an activity change are not defined. According to this point of view, cycles of different duration are principally differing from each other with respect to a number of problems. Multiplicity of different-order cycles seeming but not always taking place, for example, for 22-year cycle. Century cycle with duration of 80 to 90 years comprises 10-year components instead of 11-year ones, and it is not clear whether it is multiple to 22-year cycle. And since we are speaking about cycles and not about strictly defined periods, then the issue of multiplicity is meaningless. 11-year and 80-90-year cycles are not very similar in appearance, and it concerns the decreasing stage of periods only. The growth stage of 11-year cycle is short and the duration of that stage for 80-90-year cycle depends on cycle power. If explosion and its further decay determine 11-year cycle, then secular cycle is free, nearly harmonic oscillations. Then 11-year activity cycle is characterized by an index of frequencies, while secular cycle by indices of activity mean power. Perturbations of activity are excitations for 11-year cycles, while for secular cycle they have another significance. Taking all the mentioned into account these scientists deny subordination among cycles and the possibility of commonality for their causes and origin mechanisms.

This viewpoint was mainly formed because of weak grounding of the first viewpoint about structure and forming of the Sun system's rhythms. Despite the one-sidedness of certain problems this viewpoint is a whole and based on facts and correct definitions.

Problems of origin of the Sun activity cycles began to interest scientists together with their discovery. The German scientist Brawn in 1900 considered the effects of planet motions as the first cause, according to which periodical revolution of planet cause periodical increase and decrease in the Sun activity. Thus, according to calculations, the main responsible for 11-year periodical changes of the Sun activity, is the Jupiter, which contribution amounts 77 percent. Then in order of decrease, come Mercury with 10.1 percent, Saturn with 8.0 percent, Earth with 2.3 percent, Venus with 1.4 percent, Mars with 0.6 percent and Uranus with 0.3 percent. These "effects" are also clearly seen from other cycles of the Sun activity, which are given below.

Neptune Uran Saturn Jupiter

Frequency structure of long-term changes of the Sun activity according to Brown

165 82 84 42 30 15 12 6

Frequencies of long-term oscillations of the Sun activity detected from observation

83 42 34 15 11 6

Frequencies of long-term oscillations of the Earth climate detect from observations

85 46 34 (16) 11 (6)

But the close relation between planet's revolutions and the Sun activity long-term changes does not mean yet that the formers can force regular long-term changes in the Sun through tide waves and, moreover, when they are of magnetic nature and different scales.

Such relationship can not explain in any way a physical nature of the Sun activity long-term cycles. It simply means that the same relation correlates cycles of these two groups with the third group changes which occur in the Sun's entrails and cause them.

In 1926 Bierkness had come up with a hypothesis about cycles origin, which suggested that there were vortex rings under the sub-photosphere, but this hypothesis did not justify itself. A new hypothesis was put forward, according to which planets rather had a stimulating influence on cycle occurrence than gave rise to them, and the causes were located either inside the Sun or out-side of it, in the galaxy. The positive grain of this hypothesis was that it did not neglect planets' influence and suggested existence of free, independent force. The calculations made by scientists showed that tides caused by planets on the Sun increased its radius by approximately 1 mm which was negligible. Deviation of the Sun's of center gravity from the position of the system's center of mass is also of the same order. Taking account of the fact that the Sun activity cycles are based on changes of magnetic fields of different scales, some researchers relate their origin to magneto-hydrodynamic models. These dynamic mechanisms are basing on effects of two factors. One of the factors is rotational motions of the Sun, and the other one is existence of the convective shell where energy is periodically transferred from the Sun's entrails to the surface due to turnover of convective vortices along the normal. According to the assumptions, magnetic fields are arisen just in the convective zone.

Numerical simulations of these models on computers are quite convincing and acceptable. But phenomena composing activity are so diversified and so contradictory that scientists find if difficult to reduce them to the same denominator. Thus, conditions for origin of magnetic fields of different scales, particularly, magnetic fields of solar granules, sunspots, prominence, flakes and other formations, their transitions and interactions are not understood. According to the comments of other researchers, magnetic fields are not the only cause of the Sun activity. The problem of the united magnetic field of the Sun is not solved yet. The mechanisms, which affect polarities of magnetic fields and change them once in 22 years, are not clearly understood. Today the comprehensive kinematic and dynamic models of the Sun cycles, which may explain behavior of its activity long-term alterations, do not exist.

Thus, observed Sun cycles of all orders are related to strength and polarities of magnetic fields. But regular changes of the Sun are not restricted to magnetic nature of their origin. A huge part of cycles is occupied by stratified zonal or meridian purely mechanical, regular motions and turnovers of matter of different scales, which are related to the Sun rotation around its axis. However, nowadays-such changes of cyclic nature do not become the point of discussion by one or another reason. Then, it is obvious that cyclic changes are present not only in the visible and radio domains of the spectrum, but also in ultraviolet (UV) and X-ray domains. According to the certain sources, these changes by their total energy are at a substantial disadvantage in relation to the Sun luminosity, and, according to another sources, on the contrary they exceed it, e.g. energy emitted by thermal-nature bursts is many times higher than thermal radiation energy of bursts. Hence, in order to understand structures and forming conditions of the Sun activity cycles it is necessary to consider cycles of all natures since it is possible that cyclic changes of one nature may be converted to another one and an illusion of the problem solution will be obtained.

Thus, the problem comes again to finding out the inner structure of the Sun, laws of its motions and transformations. Astronomers have observed that more or less regular changes in stars are manifested only after substantial decrease in their chromosphere radiation comparing to their initial values, caused by their speed of rotation and magnetic fields (when a star pass over the first instant of maturation, i.e. approximately $\approx 5.10^8$ years). This is when at a particular period of star evolution a radius of convective zone comes smaller that its radius. On the other hand, according to investigations periodical changes of star's activity are related to a certain value of its mass. But it is not finally found out whether it is a general property or it is an intermediate stage in star evolution. Then arising of convective zone, according to the theory, results from collapsing stars configurations, which are arising due to gravitation condensation of diffuse matter. If it is so, that is if convection is of evolution origin, then what is the objective of its turnover in star evolution? The question and answer will be more understandable if we will try to find out the role of cycles in establishing the system's equilibrium. By which mechanisms is the Sun establishing a stability of its existence in a medium of non-stationary excitations?

Proceeding from the time-space regularities of cycles, these mechanisms may be divided into three groups: **a)** orbital motion; **b)** rotational motion; and **c)** development parameters and their changes. According to their orientations all these mechanisms have compensation nature in the Sun activity and require different energy expenses. Thus changes of the Sun orbit eccentricity, declination angle and other parameters of the orbit are caused by cosmic, or more exactly galactic, more powerful, excitations and are directed to attenuate, to dissipate their effects. Some repetitions of these parameter values in the course of time are related to cyclic recurrence of excitations. The orbital parameters possess various periods of changes, which are caused by existence of excitations with various length. Thus, each of the Sun orbital parameters dissipates by its change the energy of corresponding excitation of galactic media, till the next portion of it will come. But changes of the orbital parameter do not pass over without consequences, they particularly give rise to perturbations of rotational motions' parameters. In other words, they are excitations for rotational motion parameters. Hence in order to preserve stability the changes of rotational motion parameters should inevitably be adapted to the formers to compensate continuously a transferred momentum. However, their significance is not restricted to that only. Rotational motion parameters are also excited by other excitations of galactic medium and, therefore, the same conditions for stability require that effects of the latter be also neutralized, attenuated. Thus, changes of rotational motion parameters are affected by two mutually independent excitations, and are their dissipating agents.

Of mechanisms, the parameters determining the Sun evolution, in particular, values and sizes of convective and other dynamic regular motions occurring in it are third in importance. Through these mechanisms the energy accumulated within the Sun's entrails is periodically transferred to surface layers, redistributed and dissipated. In other words, convective and other dynamic processes neutralize and remove excess energy and strains of various nature, which arise within the Sun because of rotational motion parameters, as well as gravitational and tidal effects, thus ensuring stability of the Sun in accordance with new conditions. According to particular data, energy sources for convective turnover, together with gravitational condensation are also process of nuclear fusion, in particular, fusion of hydrogen atoms into helium. However, it is necessary, without exclusion of this opportunity, to take cosmic sources of energy into account. The Sun constantly obeys its motion and evolution laws, it permanently adjusts itself to cosmic medium, a preserving permissible values of orbital and rotational motion parameters under the influence of excitations. It would be possible if transferred energy was constantly dissipated inside the Sun in some manner, i.e. convective and other dynamic turnovers should exist. Hence perturbations of the Sun's different values and signs of the orbital and rotational motion's parameters in the course of time do not rise by themselves. But they are (compensatory) responses of the Sun to corresponding cyclic changes of galactic medium, they assure balance of interactions and the system's stability in accordance with new conditions.

CHAPTER III
GEOLOGICAL CYCLES

1. Cycles Significance

The Earth, prior to being a geological body, is without doubt a cosmic body, a permanent and stable member of the solar system. But for being a celestial body it is necessary to match and to adjust continuously parameters of its orbital and rotational motions to changing conditions of the solar system. The mentioned coordination's in the time geological axis taken as a whole determines the role of the Earth and its contribution to the establishment of the stability of the solar system. But a problem of stability in flow and hierarchic systems (as the Earth and the solar system) can not be solved so simply and clearly. Its solution usually is a multistage process of which an excitation of particular duration is compensated, but it is also a continuous process in that sense that energy of the initial excitation is compensated during the period not at once but gradually, step by step. Thus, if at the first step of establishing balance the orbital parameters of the Earth undergo changes due to excitation influence, which then give rise to similar changes of the Earth rotational parameters due to interactions. Then the latter give rise to stratified motions of different shells of the Earth at the third step of establishing balance already. Hence, changes of the Earth shells and other structural formations inside the Earth and its surface arise inevitably in response to those excitations to which they are related functionally and which are dissipated by these changes.

Thus, since some instant of time an external factor of development is replaced by an internal one, i.e. a response to excitation is replaced by a correlation mechanism of internal excitation. These mechanisms of morphogenetic correlation bond shells, structures and formations of the developing globe into one stable, but mobile set and successfully withstand various (external and internal) unfavorable effects, which can disturb the equilibrium. Despite the Earth structure, chemical, mechanical and physical characteristics of its shells and their distribution are known relatively well. Nevertheless many questions about interactions of shells remain open and not clarified. The general assumptions about contributions of shells to the division of geological labor are made.

Nature and causal and consequent connections of regular changes of various natures and scales, which are manifested especially on other (solid) shells, are much more complicated and intricate. They are elements of the final step of the stability establishing multistage process, which by their nature are adaptable, since in the opposite case they would be unnecessary and even harmful.

In this chapter we shall briefly set forth the final (so called internal) mechanisms for establishing the Earth's stability: time-dependent changes of shells and other structural formations of the Earth, which are geological cycles.

2. Cycles' Nature and Properties

The nature of geological cycles compared to the solar system's cycles is much ore multiform and interlaced. It is caused by the fact that it includes all spheres of geological life: tectonic movements, some repetitions of orogenesis, folding, sedimentation and magmatism, the ratio of surface area of land to that of water in the Earth's crust, changes in earthquake and volcanic activities, alternations of morphological formations the thermal regime of the globe, changes and oscillations of gravitation, magnetic, electric and other fields, turnover of water and air masses of different scales, etc. The mentioned geological phenomena are studied unequally and to different depths. What are geological cycles? Different aspects and parameters of phenomena served as a basis of regular geological changes. Thus, in one case, they are purely mechanical motions and oscillations of a body position; in another case, horizontal or vertical depressions and elevations (in plates) and particular arrangement of sediments; in the third case, alternations of accumulations and discharges of strains (elastic, plastic, thermal, magnetic), or sequences of morphological formations arising and demolishing, etc. At the same time, completely different problems were very important for them; in order to study these problems special methods and techniques were developed. Thus, for tectonic cycles their sizes, duration of regimes, mechanisms of origin and connections are important. These issues are the central problems of geology, because all other geological phenomena such as time intervals of forming, powers, synchronization and correlation's with other phenomena in folding processes, composition of delivered matter and sediments, age, distance from source, mechanisms of origin for sedimentation processes, and nature and composition of volcano hearth, medium and temperature of transport, regimes for volcanic contributions, are related in one or another way to solving of these problems.

Questions and problems of other geological cyclic changes are also well known. In all these problems however we are interested in origin of cycles, mechanisms of their subordination; connec-tions to regular changes of medium, etc. Diversity of physical nature of geological phenomena often hides generalities underlie these cycles and these generalities hardly are results of chances and we often neglect them.

What are the specifics of geological cycles? They, contrary to our ideas about the Sun cycles, are connected with separated shells and formations and leave their traces. These traces do not disappear immediately as do cycles observed in the atmosphere, but are "recorded" and preserved for a long time during the Earth evolution. These recordings and traces of geological cycles allow us to consider a cycle separately and as a whole or at the same time with multiple repetitions, which are impossible to carry out in the domain of the Sun cycles. Composition and structures of most of them are known due to these records. They may be studied and, hence, a succession of excitations may be restored. In the same manner, by studying changes of composition and structure of cycles in the geological time axis, it is possible to

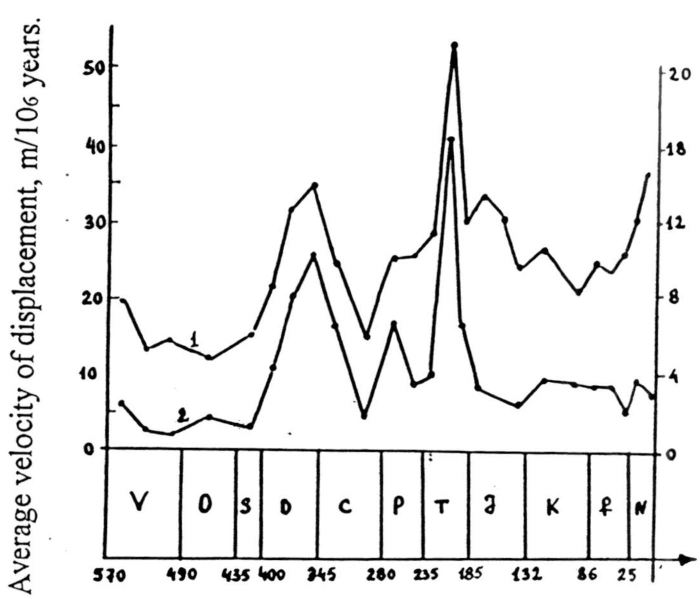

Fig. 4. *1. Average velocity of the Earth's crust oscillations, and*
2. Oscillation of magmatism manifestations (cm 10^6 years)

estimate rates of development and evolution.

What regularities of geological cycles are observed?

1. Taking account that various natural geological cycles are often observed with certain periodicity, researchers identify them with harmonic oscillations. But accurate investigations show that their notions are superficial. Geological cycles are not similar to each other in period or amplitude. In this connection Balakhovskii noted that symmetrical rhythms (harmonic oscillations) are an exception rather than the regularity, and Odesskii added that enormous experience of geological observations shows that symmetric cycles are extremely rare and they are rather non-symmetric than symmetric. This non-symmetricity of periods, i.e. inequality of growth and decrease stages (as a rule, the former is always shorter than the latter), evidences about the more complicated nature of geological phenomena. According to some authors the inequality is due to various causes. External excitations are responsible for an increasing stage, and internal properties of system are responsible for a decreasing stage. By this approach, we determine a wave nature for geological processes, and not an oscillating one.

2. It is know that geological cycles independent on frequency group do not coincide when being imposed on each other. Duration of periods is changing constantly in the course of time by unknown reason in different directions. It is a general phenomenon for cycles of all orders and natures and is accep-

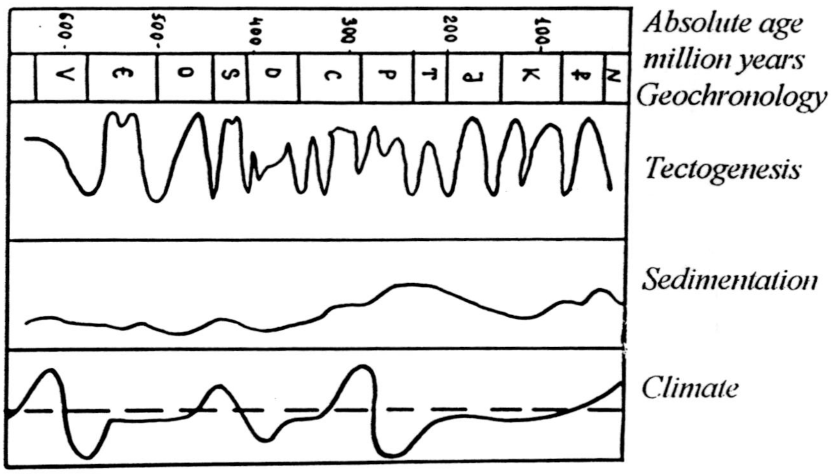

Fig.5. Behaviors of various geological processes during the last 600 million years (after Orlenok)

ted by most of geologists. But the reasons of this phenomenon do not become the subject of a special attention and are neglected or averaged in the course of investigations. At the first sight it seems that discrepancies of periods do not contain any significant information about a phenomenon behavior and they may be neglected freely. But these simple discrepancies have great significance in other connections. Particularly, they are results of neutral matching and coordination in the geological time axis of various processes occurring in the Earth entrails and on its surface, without which equilibrium between them is impossible. Establishing of stability during the cycle requires necessarily geological processes being developed with a definite rate and coordination in order to move (converse) a necessary volume of matter (quantity of energy) to a necessary place in due time. Just permanent matching of internal structures' activities and adaptability to external excitations give rise to the above-mentioned "arbitrary" discrepancies and shifts of geological phenomena's periods.

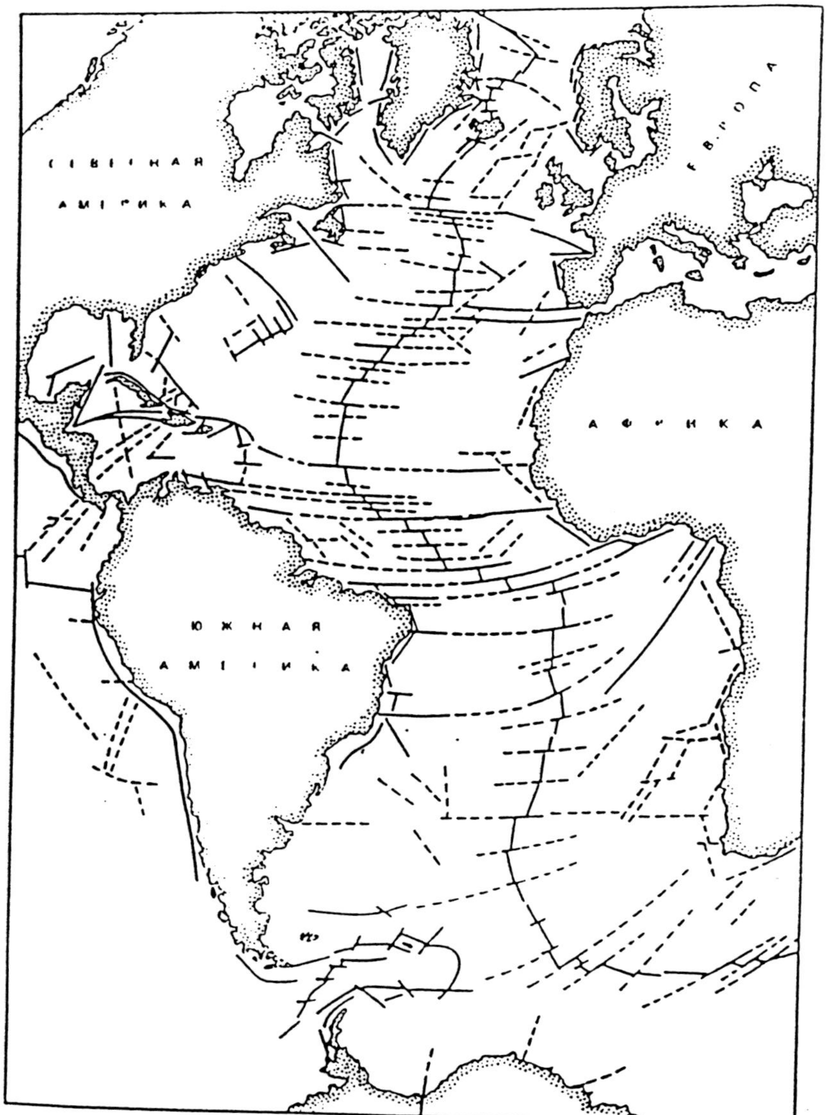

Fig. 6 *Arrangement of faults in the bottom of the Atlantic Ocean. Linear structures in the Earth's crust. (after Sholpo)*

3.Despite differences between geological cycles and theirs specify, there is a great similarity between them. Thus, according to the literature data, every cycle comprises cycles of smaller periods, and mean duration's of these cycles, as a rule, is smaller by a factor of 2 to 3 than the mean duration of the initial cycle. It is observed best of all in tectonic cycles. According to these data during Phanerozoic (the last 600 million years) the tectonic cycles have the following structure: the whole Phanerozoic is a united and entire cycle, a megacycle. It is subdivided into three large tectonic cycles lasting each 180 to 220 million years (eras), then each large tectonic cycle comprises four medium tectonic cycles each lasting 40 to 50 million years (epochs). In its turn a medium tectonic cycle is composed of 20-23 million year small tectonic cycles, which in their turn are composed of 6-7 million year smaller tectonic cycles and so on. All the mentioned tectonic cycles are similar to each other in forming and structure, the difference is, apparently in scales. Such subordination of geological cycles and cycles nested one into another is observed every-where and is accepted by geologists. Such a structure of cycles, as it was mentioned above, is also intrinsic to the solar system's cycles. It is also intrinsic to different-scale turnovers observed in the ocean and atmosphere where meaning of structure is much more evident and obvious. In these structures gradual conservation (cascade conservation), dissipation and attenuation of external excitation energy take place.

4. The fact that cyclic recurrence of some geological phenomena (earthquakes, volcanism) is not armonic or close to it evidences on its non linearity. While manifestations of earthquake and volcanism in the scale of whole planet are periodic in the course of time, then it is not an evidence that the same will be manifestations of hot point and zones of the globe. Such a notion about these phenomena is supersimplified It may be possible in the case if the crust would be homogeneous and would not develop. Then a periodicity is recognized and understood by most of geologists as nearly harmonic process, which is intrinsic to linear and homogeneous systems. Processes in the Earth's crust and entrails, probably, display such a behavior only at the first instances of the Earth forming. But permanent complications and

stratification of its structure, related to geological evolution, are ue to interactions of the Sun system and hence it is far away from meeting this definition. Finally, if a system independent on its physical nature is subjected to periodical excitations for a long time, then its all responses and reconstruction directed to establish equilibrium should inevitably include sooner or later external periods. These are exactly some reasons of irregularity of geological cycles.

3. Cycles' Classification

The axis of geological cycles is tectonic cycles to which all other regular geological changes are interlaced and from which all they are derived. In this connection it is expedient to discuss and to classify just tectonic cycles. As it was already mentioned, tectonic cyclic changes are regular mechanical motions, depressions and elevations of the Earth crust, alternations of separations between mountains and mountain valleys. Size and unit of tectonic movement are of principal importance. But this issue is not solved up to now. Some researcher (Belousov) proposes to determine tectonic movements basing on their summary results. A product of moved mass and a distance of its movement determine it, i.e. by the work done by a tectonic force. For this aim, a work spent to overcome resistance of elastic force, forces of friction and connections during deformations, emission of heat, metamorphism, quantity of spent energy and others are to be determined. But this unit of measurement is a general one, that is whether it may be used for tectonic cycles and movements (horizontal, vertical) of different orders or not, it is not clear. A direction of movement is important in that sense that those processes that are accompanied with and connected to vertical movements, are completely different from similar movements of horizontal processes. Therefore, the same force may give rise to different consequences and become a basis for incorrect interpretations. Despite these two types of movements being separated in the literature, it is in its essence unified by geologists in the problem of origin mechanism under the same name, a tectogenesis.

The longest cycles on the Earth are glaciations repeating with different periodicity. There are known six typical manifestations of glaciation in the literature. They are glaciations of lower Proterozoic aging **2,100 million** years, then upper Ripheian-**900**, Vendian- **620 to 670**, upper Ordovician-**490,** upper Palaeozoic-**310 million** years and glaciations of Quaternary epoch accompanying the human evolution. It should be noted that duration of two successive glaciations is decreasing in the course of time. Some researchers tried to justify this fact by representing this series as the following geometrical progression: **10, $10/2^1$, $10/2^2$, $10/2^n$.** The following series of numbers is corresponding to this progression: **10, 5, 2.5, 1, 0.25, 0.62, 0.156** and so on, which is often called the "magical big series" of numbers in the geological chronology. This name is caused by the fact that these numbers represent of multiples of **156 million** years. And the cycle lasting **156 million** years is considered as a galactic year, i.e. duration of one complete revolution of the solar system around the galactic center. Some of researchers try to restore the "magical offer series" of number by means of the Kepler's second law (which describes a trajectory of heavenly body in the absence of excitations), taking account of the period of the Sun revolution around the galactic center as a basis and using the astronomical estimate of today's period of the Sun revolution of approximately **212 million** years, as well as the distance from the center, and to justify a possible past way of the Sun. According to some authors, this way represents a converging Archimedian spiral in the course of time, separate turns of this special represent galactic years, which are differing from each other by their duration and geological events. The origin of time is the modern state of the system, which is considered as definitive in further restoration and justification of the turns of spiral and geological events. For these bends glaciation, tectonic huge displacements and man origin are established and substantiated. Except for duration of spiral turns, they do not differ from each other by anything. They follow a certain succession, and by no cosmogony factor you can distinguish them from each other (the evolution stages), and estimate specific features of corresponding astrophysical events. But, together with all these facts, the succession of turns or the "magical big series" of numbers is of principle importance for the Earth globe (as a geological object). Each of them represents a turning and final stage of the Earth globe activity and evolution and is distinguished by the uniqueness of geological processes, manifestations and therefore has its special name. This is evidenced by geological sedimentary and magmatic formations of corresponding ages. For the above-mentioned the "magical big series" of numbers as a whole would have a geological prototype, the event and a galactic year, according to some researchers, is to be divided into four parts according to the year's seasons. And the following series of numbers: **0.72, 156, 232, 312, 465, 620, 775, 930, 1240, 1560, 1970, 2180, 2805, 3430, 4055, 4680 million** years and so on is obtained as a result. According to some sources, completely definite structural stages of the Earth crust, metamorphism, magnetism, particular formations of orification, is corresponding to the members of the series. Some researchers change a little this series of numbers and name them differently. Anyway, neglecting their disputable aspects, an important attempt is made substantiate the numbers of this series by geological events and to restore a continuity and direction in the geological time axis.

The causes of changes in duration of the above-mentioned number series, their justifications and interpretations are of special interest. First, there is not a united opinion about cycle duration. Some researchers state that these changes are negligible, and according to others, they are substantial and have different signs. According to the data of the researchers of another group, the changes are of one sign. According to these latter's notions the cycles' duration is decreasing in the course of time and an acceleration of geological and geophysical processes or, as geologists say, the acceleration with time is observed. But there is not a united opinion as well. Some researchers attribute an acceleration of processes to galactic years only, others also add cycles with duration of **40 - 50 million** years to them, and others state that the acceleration takes place uniformly in all tectonic cycles. Magnitude of acceleration during tectonic cycles is disputable and is not final one. Some researchers consider the acceleration as constant, and others as changing one. Finally, some researchers state that the acceleration is rather restrictions of sizes, volumes and contents of geological process behavior or increase of speed, than their consequences. And hence, the ratio of adjacent cycles is different according to different authors. Often the following ratios are given: **1:3, 1:4, 1:8,** that is, each low frequency cycle comprises 3, 4 and 8 high frequency cycles. All this is more entangled when an attempt is made to clarify the causes of cycle reduction or the causes giving rise to acceleration.

What should be attributed to acceleration: speeding up of geological processes or changes of their volumes and scales, and finally, which are the definitive processes? Apparently it is difficult to answer these questions comprehensively within the frames of existing geological ideas and duration. An individual and independent method, which will be free from domination of geological fact's abundance and will use these facts correctly, is necessary for studying. Now we shall try to give such a method and to justify regularities of geological phenomenon's acceleration.

Usually in the theory of dynamic systems, we knowing a character of motion, i.e. a time dependence of **X(t)**, try to find a shape of a phase plane. In other words, we try to determine behavior of motion trajectories in the vicinities of so called singular points. Singular points are the points determining stability or unstability of a system. One and only one trajectory, to which the particular initial conditions are corresponding, passes through each point of a phase plane. Depending on choice of the initial instant and system's parameters every trajectory (movement) comes close to this or that singular point in the course of time and therefore has stable or unstable state. A stability is determined by a type of singular point, and in the case of periodic motion, by a movement of particular amplitude and period so called a limiting cycle.

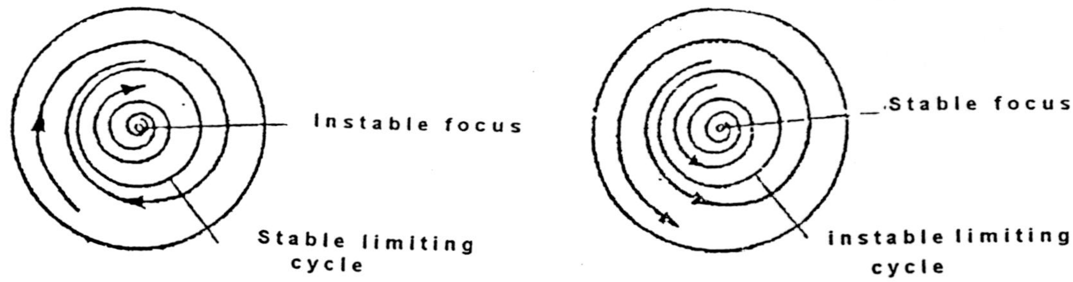

Fig.7. Characteristic types of limiting cycle

According to the theory, in the case of periodic motions a closed phase trajectory in the phase plane corresponds to the movement determined by each initial condition and, vice versa, an infinite set of periodic motions correspond to each closed trajectory, these periodic motions being differing by the choice of the origin of time.

Why is it necessary to examine trajectories in a phase plane, so called construction of trajectory patterns? The reason is that often do not knowing the equation of motion but having an arrangement of integral curves, or patterns on the phase plane, we try to say something definite about a character of motion, in dependence on time. In other words, we do not know quantitative aspects of possible motions yet, but find out in this way they qualitative characteristics.

Existence of a limiting cycle in phase pattern of dynamic systems is very important. It is a necessary and sufficient condition for existence of self-excited oscillations in a system.

What does determine a limiting cycle?

As it was mentioned above, a movement, which is described by a stable limiting cycle, is characterized by the fact that its period and amplitude are independent on initial conditions in that sense that all adjacent movements (to which an entire domain of initial values is corresponding, so called an integral domain of stability) are converging asymptotically to a periodic motion, i.e. a limiting cycle, which has a certain period and amplitude. Adjacent trajectories may come close to a limiting cycle either from outside or from inside on a spiral path in the course of time. A limiting cycle is called stable, unstable or semi-stable depending on the fact whether adjacent trajectories converge to a limiting cycle from outside and inside, or diverge, or converge from one side and diverge from the other side, or not, when $t \rightarrow \infty$. A question of limiting cycle stability is determined by studying of characteristics of linear equations with constant coefficients derived from equations of dynamic system of interest.

How can these considerations help in correct understanding of geological cycle's acceleration? It is naturally that the Earth, being a dynamic system, began its motion an the certain initial conditions and, hence, it motions should have a definite trajectory in the phase plane in the course of time. A shape of this trajectory in dependence on time in the vicinity of a singular point or a limiting cycle is not clear to us because the motion is periodic.

It is known only that geological cycles are decreasing. Can this phenomenon introduce some clarify into explanations of character of the Earth motion trajectory, or in other words, can the time shortening be a consequence of cycle shortening and hence gradual convergence to a limiting cycle? If a succession of geological causes is repeated after a certain time period as a whole, it means that the Earth, starting its motion from some point, it returns again after a certain time to the same point, i.e. its trajectory is a closed curve. And if duration of causes decreases gradually from period to period, it means that the Earth "trajectory" is not a closed curve but is a spiral. Thus it may be suggested that the Earth, preserving primarily a succession of interactions on the geological time axis, gradually "displaces" its limiting cycle, where geological events already have constant duration. These considerations are the most general ones that are given by the theory, but perhaps they are insufficient. Different sign changes of a system's parameters on the common background evidence on a more complicated mechanism of stability establishment, which will be described a little bit latter.

Problems concerning composition, beginning and end of tectonic cycles are of great interest. But these problems are not solved yet and are the topics of vivid discussion. Some researchers suggest the entire series of tectonic cycles with duration of **600-570, 220-180, 90-80, 50-45** and **23-20** million years. Other researchers assume as the tectonic cycles only particular terms of this series, and another group of researchers continues to divide the terms of this series and consider movements with higher frequencies. Main difficulties are not revealed yet, those particular forces, which are responsible for movements of this or that scale, their being external or internal ones and hereditary relations are not substituted. Besides that spatial dimensions of the above mentioned cycles are not sufficiently clear. According to the literature data differences, of the above mentioned tectonic cycles of different orders, are primarily degrees of manifestations of formations of certain types. In particular, non-similar manifestations of magmatic intrusions, folding and sediments (up to absence of their particular formations) which are characteristic for separate tectonic stages; and it's an important fact. It means that regular changes of folding, sedimentary, magmatic, earthquake, volcanic and other geological phenomena are, components and elements of separate stages of tectonic cycles, and can not occurs isoladetly. Hence the forces giving rise to them are common, which is important for consideration of the problems of their time-depended formation, alternation and coordination.

Let us find out to the which stages, the separate tectonic cycles are subdivided.

576-600 million year tectonic cycles (Megacycle). What geological phenomenon are the cycles of such order distinguished by? These cycles are distinguished very well for Phanerozoic only, and comprise grouping of three galactic years in the sense that they are not a simple repetition of events, but are connected to each other closely and are each other continuation.

What are their manifestations? According to the literature data, when passing from one galactic year to the next one the increase of mollusk formations' volumes is observed, which evidences on gradual heightening of mountain structures during the same period. The facts show that heights of mountain structures of Caledonian period (the first galactic year) do not reach mountains of Hercynian (8) period (the second galactic year), and that in their turn are lower than mountain structures of Alpine period (the third galactic year). Thus in the last **570-600 million** years a definitely pronounced tendency of mountain building process and a general growth of relief contrasts are observed. This tendency is also evidenced by the rates of growth of land and oceanic bottom relief tectonic movements, scales of intrusive and effusive magmatic activity, chemical composition of particular type sedimentary rocks, geo-synclinal areas, etc. Therefore, the unity of these geological phenomena and their succession during **570 - 600 million** years provide geologists with a basis for distinguishing a cycle of this order as the largest tectonic cycle. The cause of origin and cycle structure is vague.

180-200 million year tectonic cycle (Era). Cycles of this order are studied much more better and comprehensively and have their individual names. Every cycle of this order comprises a number of geological formations that follow each other with the certain regularity. Thus, a cycle begins usually with a rapid extensive though of separate sites of Earth surface and formation of high-thickness sedimentary accumulations, underwater basic volcanic manifestations, which are followed by folding processes, and the

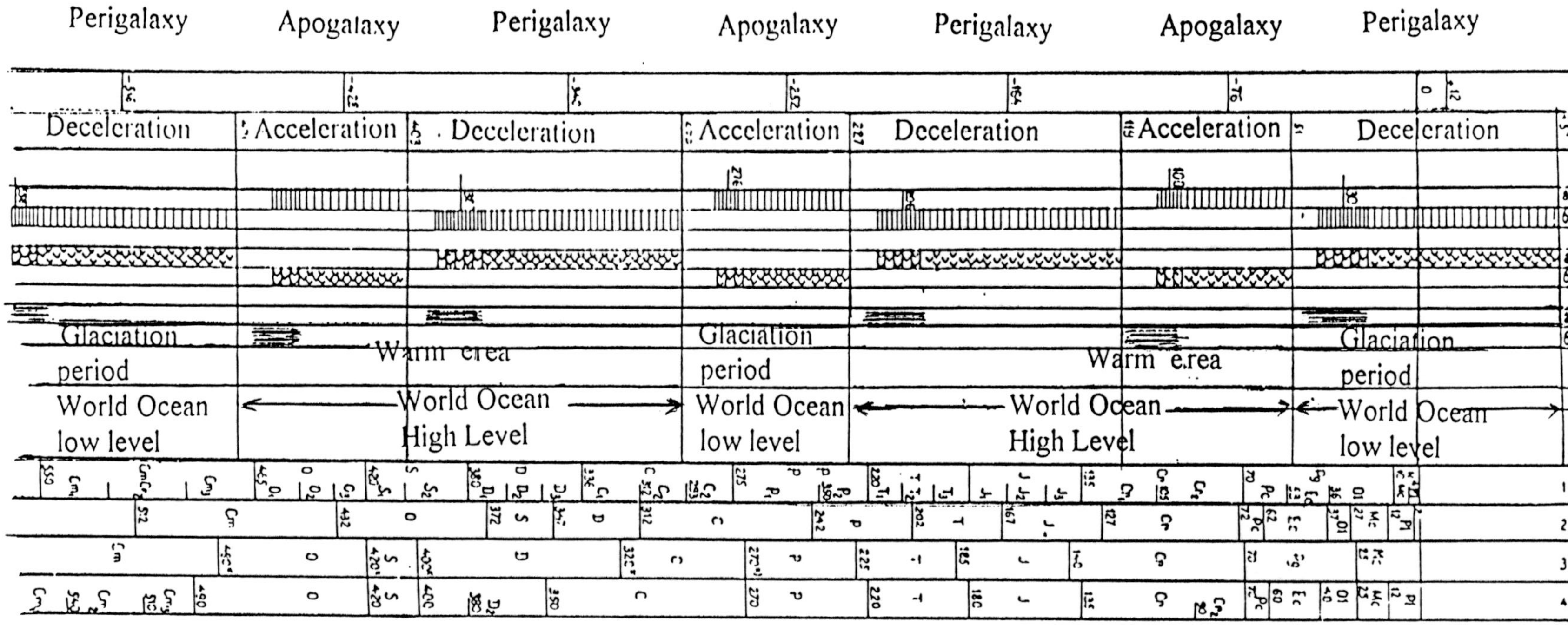

Fig.8 Superposition of changes of galactic motion velocity of the solar system, of the Earth rotation velocity, of periodical deformation of the Earth's crust, as well as a number of other galactic events on the chronological scale (after Avdeev).

I. Abnormal period of the solar system motion (duration in mln years)
II. Period of the Earth rotation velocity changes (mln years)
III Most strong deformation period (time intervals) correspond to dense cycles of the earth crust contraction deformations (folding), and slow deformation time interval to weak cycles.
IV The Earth's crust enlargement deformations; a-polar sector, b-equatorial sector,
V Alternation of sea water decrease and accumulation in the polar and equatorial regions
VI Alternations of warm and cold epochs and changes of the ocean surface level,
VII Geological chronological scale.

cycle is terminated by violent magmatism, which begins with granite intrusion contributions and terminates with formation of effusive basic thick overtrust sheets. Manifestations and alternations of the mentioned formations have a common nature and are specific to all cycles of this order and perhaps, differences are in degrees of manifestation. The most probable cause of origin is specific features of the revolution of the Solar system around the galactic.

40-50 million year tectonic cycle (Epoch). Cycles of this order are displayed in the geological time axis with different clearness. Thus, in Phanerozoic they are very distinct and clear and are transformed one in other once every **40-50 million** years, releasing approximately equal quantities of tectonic energy. Tectonic activity lasts **15 to 20 million** years (according to some sources **2 to 3 million** years) and a relatively calm stage lasts about 20 to 25 years. Cycles of this order are characterized by alternation of transgressive and regressive fancies and formations and volcanic manifestations. The unity of manifestations is as follows: at the beginning of transgression stage terrigeneous formations take place, which are followed by other not-very-thick marine sediments then bituminous, calcareous or clay sediments. Then there are followed by marine Cretaceous or clay formations, and the cycles is terminated by gypsum-, salt- and coal-bearing formations. Tectonic cycles of such order are disputed by some geologists, which take account of the fact that alternations of transgression and regression do not coincide with or are not parallel to tectonic cycles. The cause of origin is variations of gravitation force which have a period of **40-50 million** years. According to some researchers they are related to the revolution of the solar system around the galactic center, and in opinion of others, the internal forces cause them completely. According to the literature data, the structure, the cause of origin and the hereditary connections of cycles lasting **20-23 million** years and less (high frequency cycles) are more intricate and incomprehensible. Proceeding from the importance of causes of various-order tectonic cycle's origin, let us consider them in more details.

4. Generation Causes

Despite of achievements of geology in the last years, tectogenesis' mechanisms are not conclusively revealed yet. Forces, which move and regulate plates, are not understood sufficiently. Attention of geologists was always focused on the explanations of these issues, and from the beginning two directions were outlined. According to the first group of explanations the causes of tectonic movements are inside the Earth and are characterized by the specific features of its structure and activity. According to the assumptions of the second group, tectonic processes or their separate stages are related with extraterrestrial phenomena and laws. Connective turnovers and other phenomena occurring in the Earth's entrails serve as a basic for the former, and changes of the Earth rotational motion speed and rotation axis serve as a basis for the latter. Let as consider these issues a little more detail.

When discussing cycles of the solar system, we should note a stabilizing value of convective turnovers in interactions of heavenly bodies. In particular, the opinion was expressed that the energy accumulated in the body's entrails (on account of what) is permanently transferred to surface layers, redistributed and dissipated, thus reestablishing body stability in the conditions of medium interactions. Proceeding from this point of view, the suggestion that the main component force of tectogenesis is correct, but with certain assumptions. First of all, it should be noted that sometimes a contraction, oblates and gravitational stratification of the Earth, heat emitted from radioactive materials and other phenomena are advanced as a basis for tectogenesis. But they are rather secondary causes than primary ones, i.e. they are occurred as consequences, therefore are at a disadvantage according to their power. In forming and interactions of the Earth external forces, to which it should be adapted, are important. Exactly in the same manner, the convection is dominant in tectogenesis, and the Earth should be adapted to it. The above mentioned phenomena are consequences of the evolution, and their effects are temporary and secondary for tectogenesis. Then, tectonic cycles are alternation of violent and calm stages of tectonic process, hence it's natural to state that forces giving rise to this process are of the same nature, i. e. they are cyclic and have approximately the same periodicity. But usually convective turnovers realize permanent transfer of energy and not by discrete portions with certain interruption of time. It results in the fact that there exists either convection transferring energy with some interruptions, which is not excluded, or some mechanism which converts permanently transfer energy into discrete portion transfer.

The role of this mechanism may be carried out by the astenosphere, which is located on the border of upper mantle and the lithosphere and has low viscosity (high fluidity) in respect to the both. Operation of this mechanism gives rise to tectonic movements of particular periods only, and, hence, the problems of origin of other-orders' cycles again remain uncertain. If the convective mechanism would be a universal one and will act everywhere, then it is necessary to assume that it is responsible for origin of large cycles (let as say, with duration equal to a galactic year). And the other tectonic cycles are arisen from the first one through cascade transformations (similar to turbulent movements). Or, the more complicated principle of organization, i.e. a cascade of convective movements of different scales, which is responsible for different scale tectonic cycles, should be assumed. Khain considers this point of view as perfectly acceptable and convenient in discussion of causes of origin of various-scale tectonic cycles. Thus, according to his viewpoint, small tectonic cycles may be tied and small-scale convective reconstruction may take place, up to separate stages and steps of tectogenesis. But the conditions giving rise to various-scale convection are not clear yet, and especially of it evolved in a unique way. Thus according to certain sources, convective cell turnover at the initial stage of the Earth evolution was predominantly surface small-cell ones, which at the next stage became larger and deeper. The specific features of tectonic manifestation of Precambrian "pre-tectonic plates" and postcambrian, i.e. Phanerozoic "today's" plates may serve as a basis for this conclusion. If this fact would be substantiated by novel data and methods, then it may elucidate a little an evolution of convective turnovers, where is usually assumed that a radius of convective turnovers is decreasing in dependent on time, and it starts in the body entrails and propagates to surface layers and not otherwise.

Thus it seems that convective turnovers allow summarize and systematize many geological phenomena, to make clear theirs hereditary (cause and effect) relations and a direction of matter (energy) conversion. But nevertheless the convection doesn't substantiate and answer many problems of the Earth crust structure and evolution. Thus, it can't explain the regularities of faults distribution, origin and preservation of cracks and dislocations of different scales and directions (usually, 7-8 directions are accepted) in the Earth crust, movements of different scales in the equatorial and polar regions, mechanisms of small-scale tectonic displacements and other phenomena. Taking all this into account, some researchers suggest to explain tectonic movements by changes in speed of the Earth rotation around its axis and in position of the rotation axis.

According to this viewpoint, changes of the Earth angular velocity infringes the equilibrium of rotating ellipsoid, giving rise to deformation (strain) forces, which are distributed in the Earth crust in accordance with a certain regularity (meshlike). Thus is evidenced by a number of theoretical and experimental (computer-simulated) investigations. Particularly, it is confirmed that maximum - magnitude strains (elastic, plastic, thermal, agnostic) are ordered and distributed in the directions of so called inflection parallels and meridians, which can move in dependence on rotation velocity to north (or south) in the northern hemisphere, and in the opposite direction in the southern hemisphere. These forces have not time to be dissipated and attenuated and are accumulated during a cycle (rotation) and give rise, as a result, to regular tectonic movements of different nature and scales. For example, it may explain regular alternations of orogenese and folding, coincidence and adaptation of volcanic beets to the above mentioned directions. To study these cyclic changes means to study laws of the causes (strains) of their origin. But forces causing strains are numerous and diverse, hence the causes of the above mentioned geological phenomena are also numerous and diverse. According to the literature data the following forces and changes cause strains in the Earth crust : the rotation angular velocity, the position of rotation axis, the position of the Earth's core (it's located at a distance of 436 km from the geometrical center), the Earth's poles, relief contracts and water surfaces, non uniform distribution inside the Earth, possible sliding of the Earth shells relatively to each other and other forces.

Thus, regularities of strain origin, interaction and disappearance are complicated, and integrated considerations and approaches are needed. A complexity of the problems, as usual, is due to the fact that these phenomena, having different periods of changing and different rates of development and paralleling on the geological time axis, give rise to strain elastic, plastic, thermal, magnetic forces, which under the certain conditions may be transferred and reduced to each other and create an illusion of the problem solution.

Thus, according to this viewpoint, strain accumulations, which do not last long time, achieve their critical magnitudes, then are discharged and begin to be accumulated again. Particularly, exactly these alternations of elastic strains may explain the above mentioned repetitive violations of the Earth's crust intactness, origin of faults and cracks of different scales and directions, which are always accompanied by surface, medium and deep focused earthquakes. Exactly in the same way accumulations and disappearances of thermal or magnetic strains explain many problems of volcanic activity, changes of magnetic field strengths and positions of the Earth's poles (magnetic). But nevertheless, origin of tectonic cycles depending of changes of the Earth relation regime is not sufficiently understood. The point is can strain force caused by the rotation regime and accumulated in the crust give rise to tectonic movements of certain periods?

For such phenomenon to take place it is necessary to assume that there is a self oscillating system of a peculiar kind in the lithosphere, which high frequency (lasting days, years and hundreds of years) oscillations are permanently converted into low frequency (one, ten and hundred million years) oscillations. But this mechanism, if it would be found, will be responsible for tectonic oscillations of particular frequencies and therefore the problem of cycles of other orders remains open again.

On the other hand, according to the theory, the effect of rotation on convection depends on a so called dimensionless Taylor number $\mathbf{T=(2\omega d^2/\ \eta)^{\ 2}}$, where ω is a rotation angular velocity, **d** is a shell thickness, and η is a viscosity. The effect of rotation is significant when **T>>1,** if it is assumed for mantle $\mathbf{\omega=7.27\ 10^{-5}}$ rad/sec, $\mathbf{d=3\ 10^{8}}$cm and η =10^{21} puaz, then $\mathbf{T=10^{-16}}$. Hence, it may be stated that the Earth rotation practically doesn't affect the convection of mantle. But it affects the convection of external core where the viscosity is $\mathbf{\sim10^{9}}$ puaz. The mentioned feedback of rotation complicates and intricate the problem of coordination of the core and mantle convictions and a general direction of processes. According to certain sources, the radial component of strains dominates over the tangential one in the interval of distance between the center and 4200 km and visa verse in the interval from 0 to 2350 km. Thus it should be suggested that the energy of the Earth rotation is spend only for reconstruction of the upper shells, without remarkable participation of deeper processes. Hence, rotational movements are responsible for high-frequency geological cycles, and convective turnovers for low-frequency ones. It's natural that works of convection and rotation are correlated and coordinated in the course of time. Hence a question about a role of the astenosphere arises again. Two independent, on each other motions', interactions should be stabilizing. Apparently, the astenosphere plays just this regulatory role in the life of the Earth.

Let us briefly consider one else point of view concerning an explanation of regular tectonic movements. According to some researchers, the above mentioned mechanisms of the Earth's crust horizontal and vertical movements do not explain integrally the mentioned movements. In the opinion of these authors, tectonic movements are explained more comprehensively by the idea of so called the Earth's radius and volume oscillations. We are interested, as usual, only in temporal characteristics and origin cause of the phenomenon. What data and facts concerning these phenomena do exist?

At first, it should be noted that the idea of oscillation initially have not any geological ground. It was advanced only because some geological issues are explained relatively easily by means of its idea. In this sense the causes and substantiation of oscillations are completely beyond of geologists' interest. More attention was paid to these issues later, and not by geologists, taking account of specific features of revolution of the Earth (solar system) around the galaxy center. According to the literature data (Kerry, Milanovski) the cause of volume oscillations may be dynamics of the Earth's deeper layers (what forces are they caused by?), matter stratification and its re-condensation. According to this viewpoint, this process within the Earth is repeated (why?) with some periodicity and gives rise to changes of radius and volume, which causes some repetitions of geological processes. According to them, oscillations are distinctly displayed only in the most mobile sites of the Earth's crust: geosynclines and rifts. In this case, contracting compressing of the Earth are effective in the former, and extension widening forces are effective in the latter. But periods, nature, manifestations (simultaneous or successive) of these forces, their regulators and sources remain obscure and insufficiently grounded. Often it is suggested to accept as a basis for existence of these oscillations obvious changes of daily or annual coils of corals. But such changes of coral coils may be the regular effect rather of changes in rotation regime than the radius change, what is more probable. Some researchers explain oscillations of the Earth's radius and volume by influence of external forces and not of internal ones, which is more convincing from the view of point of correspondence to considerations and conclusions' scales. How is it reasoned?

The orbit of the solar system in the galaxy has an ellipsoidal form, though really it is not a closed surface. According to calculations, the eccentricity of the solar system orbit is **e=0.15** and its speed of motion in these orbit changes from **210-300** km/sec. According to the theory, the mentioned value of the speed is reached by the solar system in the most distant and closest points from the galaxy canter respectively. In other words, the solar system speed of motion during one revolution around the galactic center is smoothly changing within the indicated limits, and these changes are periodically repeating every galactic year, i.e. once every **180-200 million** years. The mentioned values of speed are comparable to the speed of light, and hence a man of body moving with this speed and its rest mass are related by the well-known expression

$$\mathbf{m=m_0(1-v^2/c^2)} \qquad (1)$$

Where **m** and $\mathbf{m_0}$ are body mass in states of motion and rest respectively, **v** is its speed and **c** is the speed of light. What will give application of this expression to the solar system? It the speed of body is $\mathbf{v_1=300}$ km/sec, then $\mathbf{m_1=1.00000055}$ and if $\mathbf{v_2=210}$ km/sec, then $\mathbf{m_2=1.0000000245}$. Hence, the increase of speed from $\mathbf{v_2}$ to $\mathbf{v_1}$ results in increase of its mass by a factor of $\mathbf{255\ 10^9\ m_0}$. The solar system during its revolution behaves as a whole and this change of mass is extended to all its members. It concerns equally to both the Earth and the Moon. Hence, the absolute value of speed and the mass of the Earth are increasing and decreasing with the same period. But these changes do not take place isolatedly and they

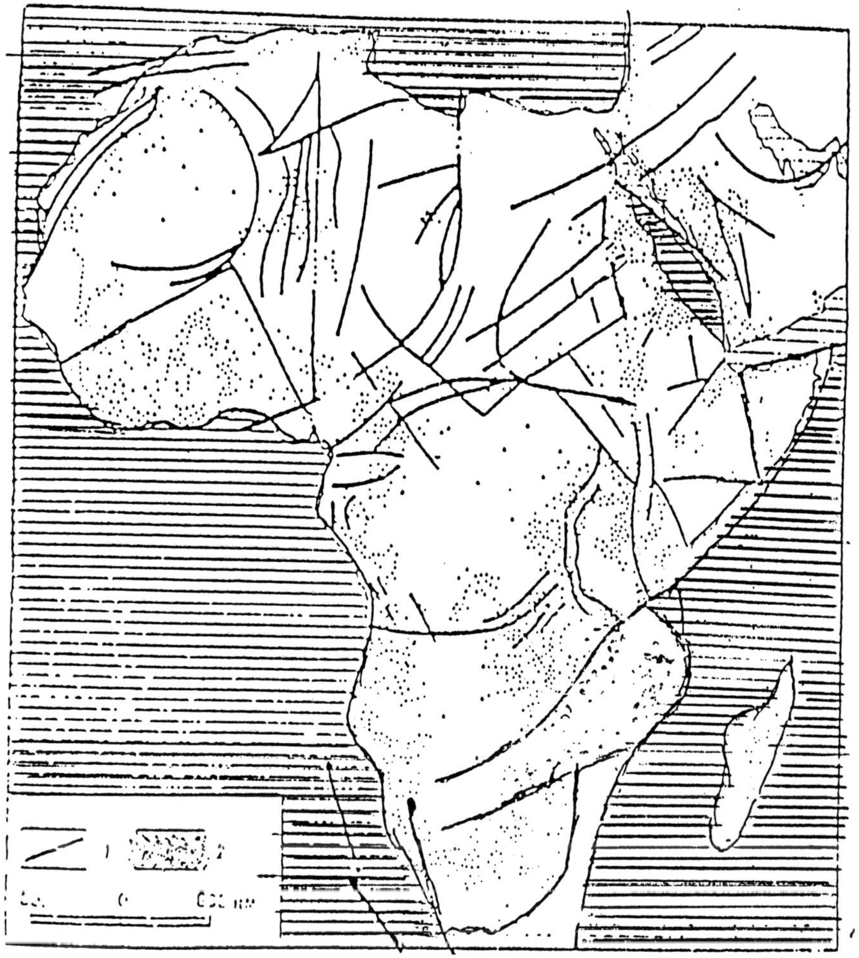

Fig. 9. *Linear structures in the Earth's crust (after Sholpo)*

Location of faults and mineral deposits in Africa. Large faults (1) Divide Africa into separate blocks, some of which are of circular Geometry. Mineral resources manifestations and formations Often coincide with circular contours (2)

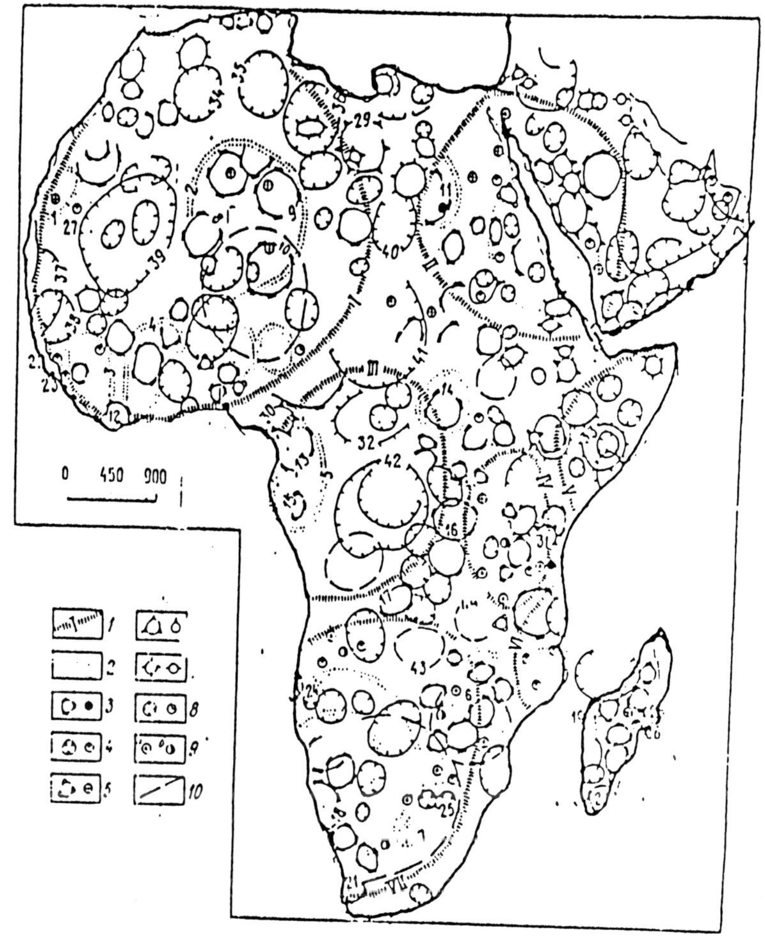

Fig. 10. *Non-linear structures in the Earth's crust (after Sholpo)*

leave traces in behavior of geological phenomena. Thus, changes of mass result, at the first place, in changes of the gravitation force inside and on the surface of the globe. Then, since the globe gravitation force is increasing beginning from the surface down to the depth of **1000** km, where it reaches its maximum value and then is gradually decreasing and is equal to zero in the vicinity of the center, and the maximum value of gravitation force is changed inside the Earth, in dependence on a magnitude of mass. Particularly, in the case of increase of its mass and gravitation force the contraction of the globe and in the case of decrease the expansions take place. In parallel to construction speed is increasing and point of gravitation force maximum is displaced toward the surface layers. During expansion of the Earth the opposite process takes place, i.e. the rotation speed and the maximum value of gravitation force are decreasing and the point of gravitation force maximum is displaced deeper, toward the globe center. Periodical redistribution of gravitation force within the Earth causes rhythmical displacements of mass. According to this viewpoint, location of gravitation force maximum closer to the Earth surface corresponds to tectonic deformation epoch and deepening location to the center to tectonic calm stages. In the first case, the solar system occupies closest to the galaxy center position in the orbit, and in the second case the most distant.

Thus, changes of the Earth motion speed during a galactic year is the main, in the opinion of some researchers, of regular tectonic changes with duration of **180-200 million** years, origin and formation of mountain building structures, geosynclines, thick sedimentary suites, folding and magmatic processes. But there is no answer for frequency tectonic movements and is not clarified which hereditary connections are between them. Then, according to some authors, activity stage is at a disadvantage in relation to calm stages by their order, but the causes giving rise to both of them are alternated with equal periodicity, therefore it should be suggested that the tectogenesis is a result of rather wave process, than oscillating (rotational) one. Finally, the speed of the Sun system motion around the galaxy can't be equally extended on all member of the system independent on its wholeness and integrity. Features of the whole (or its part) can't be extended to its parts. It is possible only in systems with linear and additive properties. It's beyond the doubt that excitations of galactic year affect the Earth globe in some way, but which way it isn't clear yet.

5. Cycles' Organization

Taking account of the mentioned structure of geological cycles researchers often brought about a question whether are regular geological phenomena a system of hereditary united and mutually connected cycles, or they are randomly grouped independent cycles with different causes of origin? Despite the simplicity of this question, apparently it is difficult to answer it comprehensively within the modern frames of geological concepts. According to the literature data, the grouping of geological cycles is a huge system of different-scale oscillating process where all regular geological changes known to us are the harmonics of one basic frequency. According to some authors, this basic frequency has long time duration, and in, the opinion of others, on the contrary, it has short period (high frequency). Developing further this concept, Baloukovski states that geological rhythms are multiple to each other and any **n**-order cycle is a cycle comprising **(n-1)**-order cycles, the latter are cycles comprising **(n-2)**-order cycles, and so on. According to him, the absolute chronological data of geology allow to suggest integer (resonant) relationships of geological cycles. In order to prove this statement, he uses the laws of orbital and rotational motions of heavenly bodies, which were discussed in the preceding sections. According to this point of view, during one megacycle a certain number of first cosmic simplest oscillations takes place, which in their turn comprise a certain number of after cycles of shorter duration, and so on. This pattern is given in Table 3.

Developing further this concept and taking account of the fact that geological cycles are nested cycles and their periods are depending on each other, it may be stated that such integer relationship should exist everywhere. And if it would be so simple and clear, then it would only be necessary to determine the primary, high- frequency process and a mechanism of frequency division. But first of all, it should be confirmed that periods are integer multiples of each other, then proceed to determination of a basic pendulum, to find out its parameters, limits of interactions, and only then to find a mechanism of frequency division. Much similar considerations are used to substantiate the other viewpoint, with the only difference that it's suggested that cycles are based on cyclic changes of long duration. Beyond the doubts this second view point is more natural and acceptable because of the following simple reason. A stability of body, independent on body's physical nature, and if a body is homogeneous, then its existence and evolution are defined by external excitation and not by internal ones. An internal excitation is a some external one that had no time yet to be balanced and dissipated within a body. External excitations are that basic causes that redefine and form the ways of evolution. Distinguishing, stratification and specification of structures within a body shall not take place independently, without external excitations and their energy. Distinguishing is caused by external excitations if they remain unchanged during a certain interval of time, hence such body will remain in this state since there are not any prerequisites and necessity to increase adaptation (stability). Interactions between body (organism) and medium in the sphere of biological rhythms is much more understandable and perceptible, than in the questions of geological cycles. In the order to confirm these considerations here too, additional commentaries and facts are necessary. A complexity of the problem is doubled here by the fact that cycles are studies rather formally, without paying serious attention to interactions of causes giving rise to corresponding physical processes and their effects. And it results in that what was said by Prof. Nalivkin: "We, stratigraphers, don't know what to do with them, where to insert them". According to Vassoevich such an approach and excessive use of the word "cycle" gave nothing to the science, but created confusion. Confusion of cause-and-effect relations of geological phenomena, their poor understanding and sometimes "absence" give researchers some chart blanch and they change development values and directions of responsible forces, as well as their duration, adapt them to requirement of separate tasks. The consideration that a strict balance of matter (energy) and effects exists within each cycle and inter actions are mutually balancing each other in time and space, establishing prerequisites for stability with respect to external excitations, are not taken into account in any way. Therefore if any conclusion is made about cycles, then it should proceed from the relations of matter (energy) conservation or balance, and not arbitrary forces.

Let us go back to a grouping of geological cycles. Without doubts, they compose a unity but not formal, according to similarity in appearance and synchronization, cascade transformations and balance relationships.

Geological cycles under the influence of non-steadiness and direction of external conditions also display similar properties and is changed slowly on the geological time axis. It's evidenced by the following geological changes, which took place during the whole Precambrian era, and were continued partly in Phanerozoic. Such as growth of the Earth's crust stratification, growth of non-homogeneity, rigidity and thickness of the Earth's crust, increase of a role of the Earth's crust structure deep faults, and growth of the gradient and amplitude of morphological structures and oscillating motions. The basis of the latter is an increase of maximum sediment thickness, which are formed during equal time intervals.

Chemical and physical evolution in the outer shells of the globe is manifested by the following phenomena: change of the atmosphere composition (decrease of CO_2 content), increase of free oxygen

Table 3.

Classification of Rhythms and Cycles (after Baloukhovski)

Rhythm, cycle	Index	Multiplicity	Duration of rhythms and cycles (mln years)			
			Holocene	Alpine 1 - 181	Hercynian 181 - 401	Caledonian 401 - 642
1	2	3	4	5	6	7
			Planet Rhythms			
Flysch rhythm	λ_1	λ_1 λ_2	0.00185	0.002	0.0024	0.00273
Precession rhythm	λ_2	$11\lambda_1$ $11\lambda_1$	0.0204	0.0216	0.0268	0.0381
Normal planet (coal rhythm)	λ_3	$45\lambda_1$ $5\lambda_2$	0.102	0.11	0.135	0.15
Mesorhythm	λ_4	$135\lambda_1$ $3\lambda_3$	0.306	0.33	0.40	0.44
Macrorhythm	λ_5	$405\lambda_1$ $3\lambda_4$	0.918	1.0	1.188	1.38
			Cosmic Cycles			
First cosmic cycle	w_1	w_1 $(15\lambda_5)$	1.38	1.59	1.78	2.00
Mesocycle	w_2	$2w_1$ $(9\lambda_3)$	2.76	3.08	3.57	4.00
Macrocycle	w_3	$6w_1$ $3w_2$	8.28	9.04	10.7	12.1
Galactic year (megacycle)	w_5	$120w_1$ $5w_4$	165.6	180.0	220.0	241
Megacycle	w_6	$720w_1$ $6w_5$	993.6	1074.0	1320	1446

quantity, change of the hydrosphere composition, reduction of the Earth's surface temperature and the atmosphere density.

Such an orientation of phenomena is also observed in changes of magmatic, metamorphic and other processes. Thus, geological evolution or history is not simple or multiple repetitions of cycles.

6. Geological Cycles' Evolution

It is natural that geological cycles did not origin at once, they had their predecessors that were replaced by new ones in following stages of the Earth evolution. It's obvious that a further evolution of the solar system will inevitably affect the present geological cycles. They will be replaced by those cycles which will be more adapted and perfect in the sense that will take account of new coming conditions as a whole. As it is evidenced by the Earth evolution and history the evolution, of regular geological changes, was developed without catastrophes, and these changes provided the permanent preservation of inter-action balance in the solar system. New cycles (say, sedimentation, transgressive-regressive, earthquake, volcanic) were formed and preserved on the background of preceding cycles (tectonic, orogenic) and continued to attenuate and to dissipate effects of new excitations without disturbing a stability state. But it should be noted that non-disturbance of equilibrium state is very conditional, and even more, here on the contrary stable equilibrium state is established due to already mentioned changes of geological cycles parameters in the response to permanent and as a rule also periodic changes of medium. For this reason an evolution of possible changes of geological cycle parameters, which at first sight, are seemed insignificant and negligible, have principle importance. Particularly, coordination and adaptation of the Earth interactions (its shells, structures, formations) to changing and complicating conditions of medium is realized and provided by these changes. It is obvious that periods of geological cycles would be unchanged then even trifle changes of medium conditions would remained negligible and unresponded, would not be dissipated and being accumulated in the course of time would be transformed to prerequisites of instability. Therefore, changes of cycle parameters are of exceptional significance in geology and especially in its evolution. As it was already mentioned, their study will allow to understand changes of the Earth globe (shells, structures, formation) equilibrium forms with change of medium conditions when passing from one stage of the evolution to another one But regularities of geological cycle parameters changes are studied poorly or unilaterally. Independent on what organizational level geological cycles are formed and preserved, they are not transformed into a simple arithmetic sum of geological cycles with simpler structures because of the mentioned changes of parameters. Their regularities are obeyed to other laws. Particularly, low-frequency (large period) cycles, being directly related to establishing the certain stability with respect to the solar system changes, play a role, at the same time, of regulators for high-frequency (short period) cycles. Thus, every time the parameters of high-frequency cycles are changed in a such way that excitation energy transferred to a particular pendulum would be dissipated and neutralized completely. In all those cases when permissible limits of parameters' changes do not satisfy external loading, e.g, new energetic interactions, then this or that stage of cycle is splitted or subdivided. Thus, a necessity for a stage to be subdivided is put forward when an initial cycle is unable to dissipate changed requirements of medium (internal or external excitation). Therefore stability necessities to establish such a structure of regular geological changes, which will accurately reflect changes in energy and matter, though such subdivisions. These way results in establishing of strictly subordinated new geological cycles within a time, which have higher frequencies and play a role of cycle servicing, promoting the initial cycle. They already do not interact directly with external medium changes and become more independent on them. These internal cycles start to operate independently without external signal and come to life at necessary instants

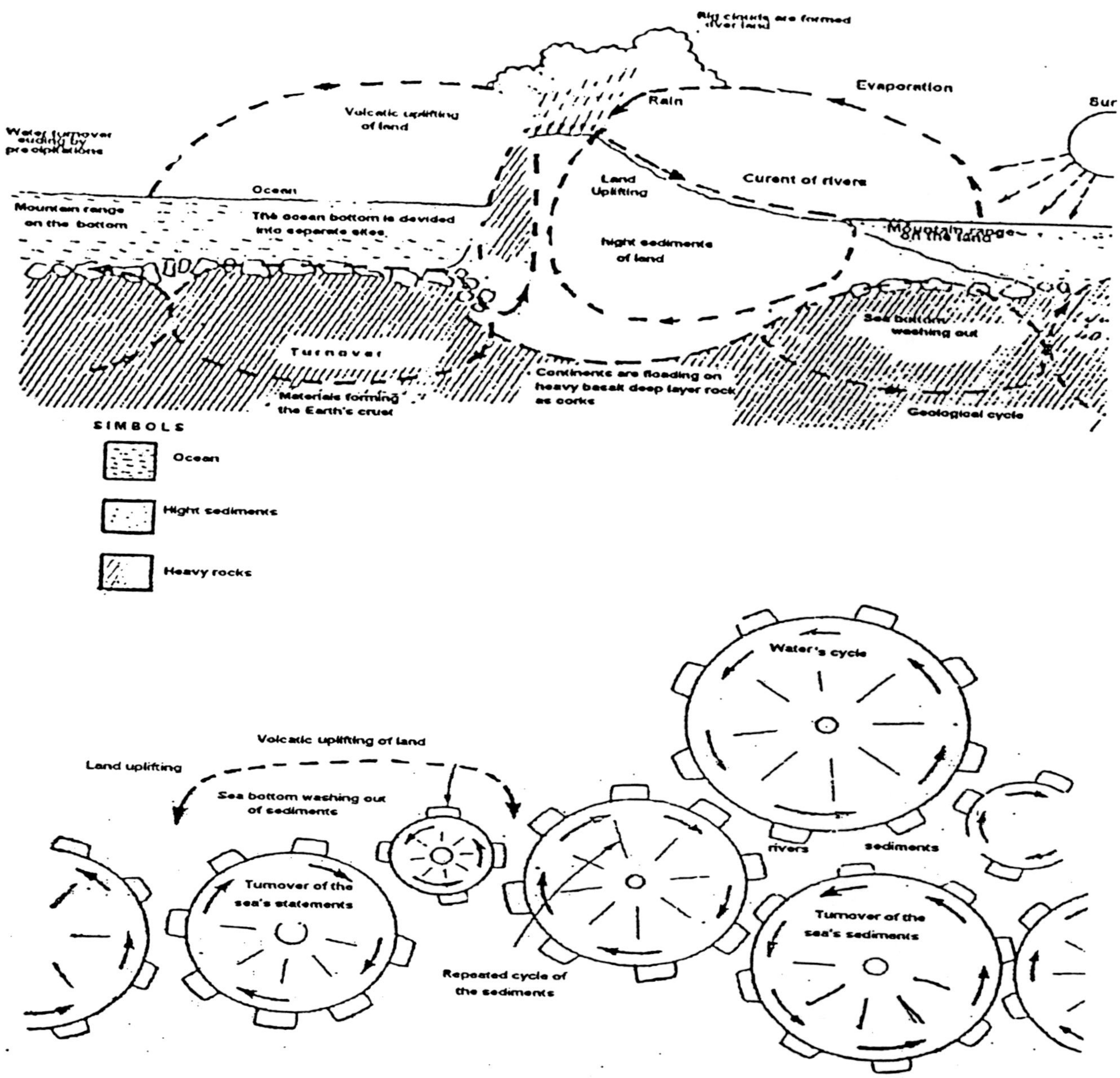

Fig. 11 *Simplified scheme of geological cycles. They include cyclic changes of the oceanic bottom sedimentation processes (rock formation, orogenesis, volcanism) and water cyclic changes, which form a united system (a)*
A similarity between geological cycles and years. All years are engaged to each other and put each into motion, carrying out its allocated work share in total work (b) (after Odoum)

of time, and it may be caused by some external signal (let us say, some tidal wave or 11-year cycle). Thus, during the Earth evolution, in the complicated inter, relations of conditions of the solar (galactic) system interactions, a chain of geological cycles interactions is formed in the globe, where frequencies of new geological cycles are continuously increasing (and periods are decreasing), become more and more independent on medium changes and service preceding cycle of a longer period.

Then, the Earth globe begins spontaneously be to be subjected (come into resonance) to influence of corresponding frequency changes of the solar system, which result in forming and preserving endogenous higher-frequency cycles, which previously were not interacting. As a result, a sphere of interactions between the globe and the solar system is enlarging gradually and stability becomes more and, more comprehensive.

7. Cycles' Search Methods

A simulation of cycles is a difficult task and we will outline general ways. Each of the Earth's shells is a unity (a system). It means that its response to an excitation is an integral one and is repeated every time when an external excitation is repeated in the same way. According to Demenitskaya in the spectrum of natural processes quite distinguishable maximal of their progress, which are corresponding to different phenomena proceeding in different shells of the Earth, are observed. In other words, quite definite processes with dominant periods characterize every shell. Their equilibrium states depending on physical properties (particularly, viscosity (η), stiffness (μ) are restored with different rates. Value η and μ being sufficiently lower in the hydrosphere and the atmosphere than in solid shells due to reconstruction and permanent turnovers force an excitation to be dissipated faster, and an equilibrium state is restored during the cycle. Exactly in the same manner a change of rotation axis position, cause quick restoration of the equilibrium's state of these two shells. An excitation has no time to be accumulated within them, and dissipated permanently. In the Earth's crust, the upper shell, where relaxation with respect to the Earth rotation speed is very high, the excitation effects have no time to be dissipated a cycle and begin to be accumulated. Just in the same manner, when change in the rotation axis position is not subject to resistance from corresponding fast motion and reconstruction of the shell, it's accumulated again, causing strains (elastic, plastic, thermal, magnetic). General laws of strain discharge and removal are manifested in the Earth's crust by faults of different scales and directions and earthquakes, volcanic activity, folding and other geological cycles. An excitation in the mantle is dissipated through convective turnovers, which become internal sources for strains in the crust. What is the role of the core in the Earth life is not clear yet (what transformations are made try it), but it is obvious that it , similarly to other shells, is under the influence of external excitation and transforms and transfers the excitation energy to upper shells. Thus, the Earth's shells, in the conditions of external excitation influence and due to them, have certain inner structures in the geological time axis and are steady or periodical repetitions of a number of simple structures, which conduct dissipation. Powerful and long-duration excitations affect inner shells, and weak and short-duration ones affect the crust, the hydrosphere, the atmosphere and their separate formations. This is the reason why the earlier are considered as all-planetary cycles and the second are regional and local cycles.

This subordination of interactions may be explained by means of the non-linear oscillation theory. Thus, usually a frequency ω of a system (a rate of restoring process) and a frequency ω_1 of external excitation may have the following relations:

Table 4.

Characteristic / Shells	Depth of layer, km	Viscosity din/cm^2 sec	Stiffness, din/cm^2	Relaxation, sec	Density, g/cm^3	Pressure, k bar	Free fall acceleration, cm/sec^2
Atmosphere	-16 - 0	1.8 10^5			0.95 10^{-3}	0.4	
Hydrosphere	0 - 4	0.1 10^2			1		
Crust	0 - 33	9.10^{22}	0.3 10^{12}	3 10^{11}	2.7 - 3.0	0	0.82
Mantle(astenosphere)	A 33 - 300				3.0 - 3.6	0.1 10^6	1025 -1026
upper mantle	B 300 - 1000	10^{24}	3 10^{12}	10^{12}	3.6 - 4.7	0.3 - 0.8 10^6	1026 -1032
lower mantle	C 1000 - 2900				4.7 - 5.7	0.8 - 1.8 10^6	1032 - 850
Upper core	D 200 - 4980	10^9	3 10^{10}	3 10^2	9.4 -11.5	1.8 - 3.1 10^6	850 - 380
Lower core	E 4980 - 6370		2 10^{12}		11.5 - 12.5	3.1 - 3.2 10^6	380 - 0

1) ω_1 >>ω ;2) ω_1 << ω; and 3) ω_1 ~ ω. In the first two cases (when $\omega_1 \neq m/n\ \omega$) a system practically doesn't respond to external influence, it as if doesn't exist for the system and a system's oscillation frequency is defined by its own frequency (determined by its internal properties, particularly, mass, relaxation, etc.). But if ω_1~ ω which is the condition of resonance, then a system begins to interact with an excitation and a resulting oscillation takes place at the frequency that is close to an excitation frequency. The system own frequency is completely inhibited and is not manifested. Then, if a system (necessarily of a hierarchic structure) interacting with different-frequency excitation (particularly, $\omega_I < \omega_{I+1}$, (i=1,2,..,k) preserved its stability in the geological time axis, it means that oscillation frequencies of its components are adapted to and reflect external frequencies, i. e. are also subordinated and directed to dissipate their effects. Which is the parameter definitive for these frequency subordination? Usually, frequency of non-linear oscillation is proportional to a mass (or relaxation) value. Hence, if component masses (or relaxation) of a system are selected in due manner, then it's possible that a sys-tem would exist under the above mentioned excitation conditions, therefore it would be stable. But what are the characteristics of the Earth's shells? Some of them are given in Table 6.

The table shows that distributions of shell masses (relaxation) display essentially the mentioned tendency. That is, periods of dissipate processes occurring in shells, depending on their viscosity and stiffness, are growing. According to some sources, these connections in biorhythmic scale are expressed by a simpler relation $T=kq^{2/5}$, where **k** is a proportionality coefficient and is equal to **3 10^7 cm$^{2/3}$g$^{-2/5}$ sec$^{7/5}$. It** is manifested in a huge domain of frequencies (approximately 4 0 orders) and this is a reason that it is considered as a general law.

It is possible to try to derive simplest relations of energy conversions' balancing taking place in the Earth's shells. Usually when studying uniform systems the relations of matter (energy) conversation and preservation balance are rather important than its dynamic equations. In such systems the dynamic emerges in the result of a balancing energetic conversations. Hence when passing from one cycle to another one, it may, depending on excitation amplitude, be displayed in different forms and may not be a characteristic one. Let us consider the following set of equations:

$$dx_1/dt_1=a_{11}yx_1-a_{12}x_1-a_{13}x_1x_2$$
$$dx_i/dt_i=a_{i-1}x_{i-1}xi-a_{12}x_i-a_{13}xix_{i+1}+v_i \qquad (2)$$
$$dx_n/dt_n=a_{n1}x_{n-1}x_n-a_{n2}x_n+v_n \qquad i=1,2,3,...,n.$$

where **$x_1(t_1)$, $x_2(t_2)$,...,$x_n(t_n)$** are functions of different periods, and **t_1 > t_2 >...> t_n** are periods, a_{ij} are constants, **v_i (i=1,2,...,n)** are excitements, which have the same frequency as **$x_i(t_i)$** (i=1,2,...,n). The set of equations (1) have many possibilities. Attributing various meanings and contents to **$x_i(t_i)$** functions, it is possible to study problems of different natures, particularly, behavior and interactions under the conditions of external stability v_i **(i=1,2,...,n)** of object of non uniform, hierarchic structures. If **$x_i(t_i)$** functions **(i=1,2,...,n)** are energy quantities in the **i**-the level of the system, then these equations will reflect a balance of energetic conversations and chain which will be realized by the system under the conditions of given excitements $v_1,v_2,v_3,...,v_n$. If the levels mean the globe shells, then the mentioned set of equations will represent the simplest activity of the "Earth's globe" (interactions, energetic conversations, etc.) under the conditions of given excitements $v_1,v_2,v_3,...,v_n$. What is a sequence of energy (matter) conversations and its dissipation in this case? Let us try to find this direction. The first of the coefficients **a_{11}y** is that quantity of energy which is received by core from outside through interaction with a low-frequency excitation. This portion of energy is dissipated due to the core activity within the time interval t_1 till the next portion will come. The dissipation is expressed by the coefficient a_{i2} and realized by means of interval structures activity.

What does it mean and how does it take place? Let us refer to analogies: in physiological rhythms about 80 to 84 percent of the excitation energy is dissipated, in ecological rhythms about 60 to 80 percent, and in atmosphere cyclic turnover about 90 to 94 percent. Remaining portion of energy is spent within a system, it is used to preserve an internal regularity, for phase transitions and other processes, which realize this dissipation. The cause of slow and directed changes of physical-and-mechanical (plasticity, excitability, etc.) properties of a system is phase transitions, which are also the cause that a system never returns to its initial state after excitement. Now it is not difficult to find that something like that phenomena should take place in the Earth's shells during neutralization and dissipation of an excitation. Perhaps, it should be suggested that dissipation of excitation energy in shells amounts about 70 percent, and the remaining portion is spend for preservation of its structural regularity, inevitable changing of viscosity, stiffness and other parameters. The indicated dissipation value in shells may make clear the problems of the cyclic changes of the globe thermal regime and its origin, which are disputed by geologists now and then. But let us return to justification and commentaries of the equation coefficients.

Thus the coefficient a_{i2} is an integral characteristic of energy expenses of all kinds (dissipation, kinetic energy, phase transformations), which are arisen when the stability is re-established in a shell. When interacting, shells transfer to each other certain portions of energy, which are described by the coefficient a_{i3}. For simplicity, the interaction function is taken as $x_i x_i+1$. The next shell, which has other period of oscillation (or restoration processes) $t_2<t_1$ is subjected now to two excitements. The first function with the underlying shell and the second one is the external source f_1. These portion of energy are dissipated during the cycle (of t_2 period) because of the shell activity, i. e. reconstruction and turnover, and a stability within the limits of shell is re-established till the end of cycle. The same is repeated within the limits of the third shell, but the period being $t_3<t_2$, and portions of energy transferred because of interaction with the second shell and from the external source f_2 are dissipated. This process, i. e. a process of neutralization and dissipation of low-frequency's excitement, is continued up to the upper shells of the globe, the hydrosphere and the atmosphere, where its energy dissipation is completed and the stability of the whole system is re-established. The process is repeated in the case of new excitation of the core, i.e. during its next period (cycles and so on). Without doubt, such explanation of and general approach to the Earth globe "activity" is far being sufficient. But inclusion of its all shells into such united transformation chain is useful in the sense that it allows to outline the frames of their activities and roles in establishing the system stability. Without clarification of these problems the questions of the globe structure and organization, the crust origin and evolution, and a number of other questions remain unclear. When a role of shells would be clarified (here the questions concerning it are only stated), then it will be necessary to find out a total amount of energy for each shell (internal, potential and kinetic energy), possible conversation the nature, and dissipation of shell interactions and mechanisms, which carry them out, to estimate coefficients a_{ij}, and to try to formulate a question of the system stability, proceeding from characteristics of $v_1, v_2, \ldots, v_n$ excitements.

In order to understand more precisely internal structures of lower shells, energetic cyclic conversations, reconstruction and turnovers, let us make clearer interrelations of the Earth globe shells and cosmic excitations.

3.2 Relation between Cosmic and Geological Cycles'

The analysis of different scales of geodynamic regular displacements and motions has significance for the generation of forces their nature. Some researchers consider those forces as endogenous, others as exogenous, attempting to explain them by objective law and by the development of the internal or external (cosmic) environment. With that connection appropriate facts are deduced, whose acceptance or rejection lead to different geological conclusions, deductions and inferences. The structure and function of the earth's depth and shells, seemingly, are not possible to explain only by the geophysical concepts and means. In recent years, interest has increased in the synergetic interpretation of existing geological phenomena and motions, the distribution and allocation of the physical-mechanical parameters and descriptions. In this section an attempt to present such an examination.

1. Shells' Relaxation Times

It is well known that the linear dimensions of the Earth do not exceed **L = 7x10^9 cm**. On a sphere of similar dimensions, based on the velocities of propagation of longitudinal and transverse waves, one can separate and distinguish several shells. In this manner, Bullen (1975), using Jeffrey's data, presents the following distributions of wave velocity, density, pressure, viscosity and other parameters (Jeffrey, 1959).

Table 5.

Zone	Depth **km**	**r** g/cm^3	**p** dine-/cm^2	**k** 10^{11} N\m^2	**g** m/-sec^2	**a** km-/sec	**b** km/sec
	0	2.94	0	0.38	9.82	6.32	3.56
A	33	3.32	0.009		9.85	6.75	4.36
B	413	3.84	0.141	1.731.	9.86	8.97	-
C	924	4.35	0.379	3.493.	9.98	11.42	6.366
D	2000 2898	5.11 5.56	0. 1.360	5.10 6.10	10.01 10.73	2.79 13.64	6.92 7.8
E	2898 4000	9.80 11.42	1.360 2.470	6.55 10.33	10.73 7.87	8.7 9.81	0.0 0.0
F	4980 5020	12.17 12.25	3.300 3.280	13.26 -	6.78 4.31	10.44 9.40	0.0 0.0
G	6371	12.57	3.610	-	-	14.16	0.0

The shells have different physical-mechanical properties and chemical compositions. In particular, they are distinguished by different densities ρ, pressures **p**, viscosities η, shear moduli μ, temperatures **t**, and a number of other parameters. Usually η is considered to be internal friction of the medium, and it is defined by relating the tension and the rate of deformation:$\eta = \sigma \mathbf{de/dt}$. A similar relation when the rate of deformation is replaced by the deformation itself, defines the hardness: $\mu = \sigma$ **e**. From this relationship, it follows that the measures of viscosity and hardness differ by a multiple of time dimension:$\eta=\tau\ \mu$. The quantity τ represents the relaxation time of the medium and has great significance in various aspects. In phenomena of long duration, say 10^2 years, elastic behaviors (movements) in the earth's crust become unacceptable, and are substituted by viscosity or by plastic behavior (movement). Consequently, τ equals transition from elastic to viscous behavior. The determination of τ is also important for the establishment of equilibrium and balance. Actually, relaxation represents the interval of time necessary to restore the system. In other words, τ is the duration of the reconstruction, rebuilding processes which arise in the system (medium) after every action. The determination of η, μ, τ is a complicated problem for each shell. Usually, for determination of η, μ, τ processes with evidently long duration are used, specifically, transgression and regression, slow raising and reducing of separate plates in the earth's crust. In the last case for example, if the velocity of the earth's crust plate is equal to 1cm/year; then, according to (Orlenok, 1980), $\eta=9\text{x}10^{22}$ puaz. According to Jeffrey's data for the external core, the value of η is 10^9 puaz, for the upper mantle $\eta=9.10^{11}$ etc. The known formula of Pray for the determination of μ reads:

$$\mu = 16(1-0{,}03r^2)10^{11}\ \text{dine/cm}^2 \qquad (1)$$

where **r** represents the radius. Eq.1 is not useable, because it does not include spasmodic alteration. Currently, the distribution of η, μ, τ by shells has the following form (Table 6.) As it follows from the Table 6, there are two values of τ for the core: ∞ and **1**. However, both of them are unrealistic from the geophysical standpoint. In the first case, the core represents an absolute rigid solid; and, in the second, a Newtonian liquid

Table 6.

	dine/cm sec	μ dine/cm	τ sec.	T year
Earth aver	10^{20}	10^{12}	10^8	10
Lithosphere	$9\text{x}10^{22}$	$0.3\text{x}10^{12}$	3.10^{10}	10^3
Astenosphere	$9\text{x}10^{21}$	$0.7\text{x}10^{12}$	10^9	10^2
Mantle	10^{24}	$3\text{x}10^{12}$	10^{12}	10^5
Upper core	10^9	$0 - 10^9$	∞ , 1	∞ , 0

which instantly reacts to an influence or coercion. Hence evaluation of the core **t** should be accompanied by other (for example, synergetic) presentations.

2. Shells Parameters' Hierarchy

Earth as a hierarchical system has functioned for a long time in interaction with the environment. Realizing transformation dissipation forces and coercion of energy (angular momentum, impulse and other kinds of interaction) requires that the base parameters of the structure are subjugated to the self-hierarchy.

Definition: *We should say that the embedded media (spheres) interacting in the space have self-hierarchy structure, if their characteristics are obtained from one another by similarity transformation.*

The self-hierarchy of structures and parameters are prevalent in the astrodynamic, geologic, biological other developing systems where they pursue certain goals (Alphven, and Arrenius,1975; Tides and Resonance,1975). Also in the geology, more than one expert has noticed the hierarchy of the geodynamical structures (Ferhugen,1972; Kosigin,1972; Sadovsky, 1986). Questions arise as to whether or not the law of self-hierarchy is applicable to the shells, and if it is, in what form it appears. In ascertaining geological peculiarities of the shells, depending upon the distribution of **p, ρ, η, μ, g** and other parameters, it is impossible to ignore or disregard the transfer of angular momentum and impulse, and the forms of their conversion and dissipation (loss of energy due to the rub, viscosity, resistance). Motions are stipulated by external sources of energy. Shells continually balance the influence of external energy on the geological scale of time. The principle of self-hierarchy signifies that the stratification of the depth of the Earth as a result of the division of the planet's effort (labor) occurs in a consistent manner. Each shell, due to its own dimension, density, relaxation and other characteristics, is adapted to the influence of definite extensions. The shell partly pumps, converts and dissipates above mentioned influences and, in this way, fulfills its quota (share) contribution in the establishment of the equilibrium of the system. Therefore, if μ_i and η_I, i= 1,2,3 represent respectively the magnitudes of hardness and viscosity of the lithosphere, mantle and core respectively, then self-hierarchy through for them means;

$$\mu_1/\eta_1 > \mu_2/\eta_2 > \mu_3/\eta_3 = q < 1 \qquad (2)$$

What facts exist to justify (2)?

1. Alterations of μ_i and η_i in radial direction occur in parallel with some deviations (Orlenok,1980).
2. Orientation of the geophysical and geodynamic processes (Khain, 1964).
3. Existence of regular behavior in the interactions: Accidental values of shell parameters and characteristics could not form cyclic, monotone motions (Barenblatt 1982; Sarafyan, 1990).

According to (Demenistkaya,1981), processes' lengths, proceeding in the shells, are linked upon their substance viscosity and are described, in the bilogaritmhic scale by the relation $T = Kh^{3/5}$, $K = 3x10^7$ $cm^{2/5}$ $g^{-2/3}$ $c^{7/5}$. So, the magnitude of the core's **t** must be $\infty >> \tau >> 0$. Therefore, supposedlyedly, any geological cycle is not symmetrical. Increasing phases are always shorter than decreasing phases. The calculation and measurement of the tectonic cycles show tectonic active and tectonic passive period form respectively **30-35%** and **70-65%** of their whole length. The biggest tectonic cycle, in other words galactic cycle, generates the most powerful and the slowest alterations in the Earth. In particular, in the first place internal core. Considering the self-hierarchy character of shells' parameters and of the passive phase the biggest tectonic cycle, the value of the core relaxation time should be as $1.2x10^7$ years. Thus the Earth, as a complex system, possesses the same peculiarities as the above-mentioned system. It transforms and dissipates the energy of external sources through its shells. Shells with long (short) relaxation times are under the constant influence of forces with long (short) duration and equilibrate them.

3. Exogenous and Endogenous Forces

The nature and characteristics of forces and influences, external with respect to the given shell, possess significant importance. Some experts find those, within the solar system, influences of long duration are absent (Frank-Kamenesty, 1964). In these case forces of interaction, arising due to repeated mutual positions of celestial bodies, are small and have high frequencies. Therefore, they can not serve as a source for the mentioned feature. However, one should not over-look the possibility that the direction of pumping energy transmitted in a similar manner is accumulated and dissipated by low frequencies. Otherwise, the trend of the pumping energy, which is observed in the geological cycles, has a tendency to slip from high frequencies (at small scales) to low frequencies (to large scales) (Sholpo,1980; Sarafyan,1985). Phenomena like these are well known in hydrodynamics and geophysics in the presence of negative viscosity of media (Starr, 1980; Lorenz, 1973). However, facts confirming similar interactions carry particular character. Therefore, according to (Belousov, 1975), there are no energy sources with a long length in space that could generate tectonic and other phenomena of long duration observed on the Earth. Only applicable force could be the variation of the Earth's gravitational field. But it is negligible compared to the gravitational field itself. Moreover, in such cases, the Earth appears as an absolutely solid body. Therefore, the governing motion becomes the overall motion of the Earth, but not the geo-dynamic or geological motion. However, the magnitude of the gravitational field of the Earth and the combined forces of interaction arising in certain mutual positions of celestial bodies (i.e. at the syzygies, and phases) might be comparable to or sometimes greater than their average value. The appearance and disappearance of tidal forces confirm this fact. According to Kropotkin (1970) and Khain (1973), the geophysical mechanisms of Earth are based on the principle that the convective motions in the core generate physical-mechanical processes in the mantle, i.e. convection of the second order. The latter in turn generate rheologic, geo-dynamic and geological processes in the lithosphere and crust. All these different scale processes yield to each other by the magnitude of forces. Namely, their spatial and temporal scales are evinced in self-hierarchical order. Such a presentation is complete enough to overcome most contradictions of the geophysics and geo-dynamics. Its vulnerability is only in our ignorance of the external source of energy generating convection in the core. There are some attempts to explain the convection but they are not convincing. For instance, in the references (Elasser, 1950; Runcorn, 1954; Urvey, 1952) it is assumed that:

a. In the core there is thermal convection. However, the thermal convection flow is generated only as a result of the convection motion and not conversely. Even if it was the case, the source of energy generating thermal convective flows in the core remains indefinite.

b. There are small differences in the chemical composition of various parts of the core. However, small heterogeneouses are able to generate powerful, constantly acting motions in the case of negative viscosity. But the motions would be turbulent rather than convective.

c. There is incomplete equilibrium of the core. First, neither forces nor the conditions in which the equilibrium can generate active convection are specified. Usually the convection itself, due to the permanent mass and heat transfer, leads to equilibrium of the system with respect to excitations, but not vice versa.

Gutenberg (1959) pointed out that reliable sources of energy that keep the convection in the core have not yet been found. Some experts, based on the presence of similar processes in the celestial bodies, consider the existence of convection in the core as definite. They make their conclusion, proceeding from the existing and relating it to the internal forces and processes, independent of the external conditions. It should be pointed out that the convection takes place in a certain phase of object-medium interaction and represents some form of transformation and transfer of energy. Convective motions yield to other types of motion if external conditions are suitable for that. Convection, for instance, is converted to turbulence (diffusion) if the source of energy is extended (suppressed). Theoretical and experimental investigations confirm this (Hayashi, 1966; Hydrodynamic Instability, 1981). The basic conditions, for the generation of convection, are following 1. Critical value of the influence 2. Definiteness of the system' dimensions (in this case, the radius of Earth). Thus, the convection can not be generated, exist for a long time, and disappear spontaneously. In order to have convection, corresponding sources, influences, and conditions are necessary. One would think that the system's endogenous forces, at the moment **T**, are the same exogenous forces, which entering into the system at the previous moments of time (**T – i**) and do not converted and balanced completely into internal forces. Otherwise, the reacting mechanism (respond to the external excitation) is not transformed completely into correlating mechanism (respond to the internal excitation). So, in a certain sense, exogenous and endogenous forces are the same. Based on such view, it is necessary to assume that there are definite exogenous forces, which generate convection in the mantle, and the predominant period **T** is greater than the relaxation time which, is on the order of 10^6 years. The source of that force decisively is, peculiarities in the revolution of the solar system around the center of the galaxy. The galaxy's center and the center of mass of the galaxy can come close together or reach of the distance of more than several galaxy center's radii. Between these two extremes, the galaxy's orbital angular momentum increases or decreases, provoking appropriate long-length cycles if its constellations particularly the solar system's galactic year **(180-200 million years)** the galactic half year cycle **(90-100 million years)**, the galactic season **(45-50 million years)** and so on. In Fig. 12a the long-length's of some of the geological cycles during the last 660 million years are shown. Similar connections and interactions of much small scales and potencies, occur also in the relationship between the solar system's center of mass and the sun's center, and the between Earth's center of mass and its geometrical center. The lengths of both oscillations are much shorter than the length of the galaxy's center oscillation. In Fig

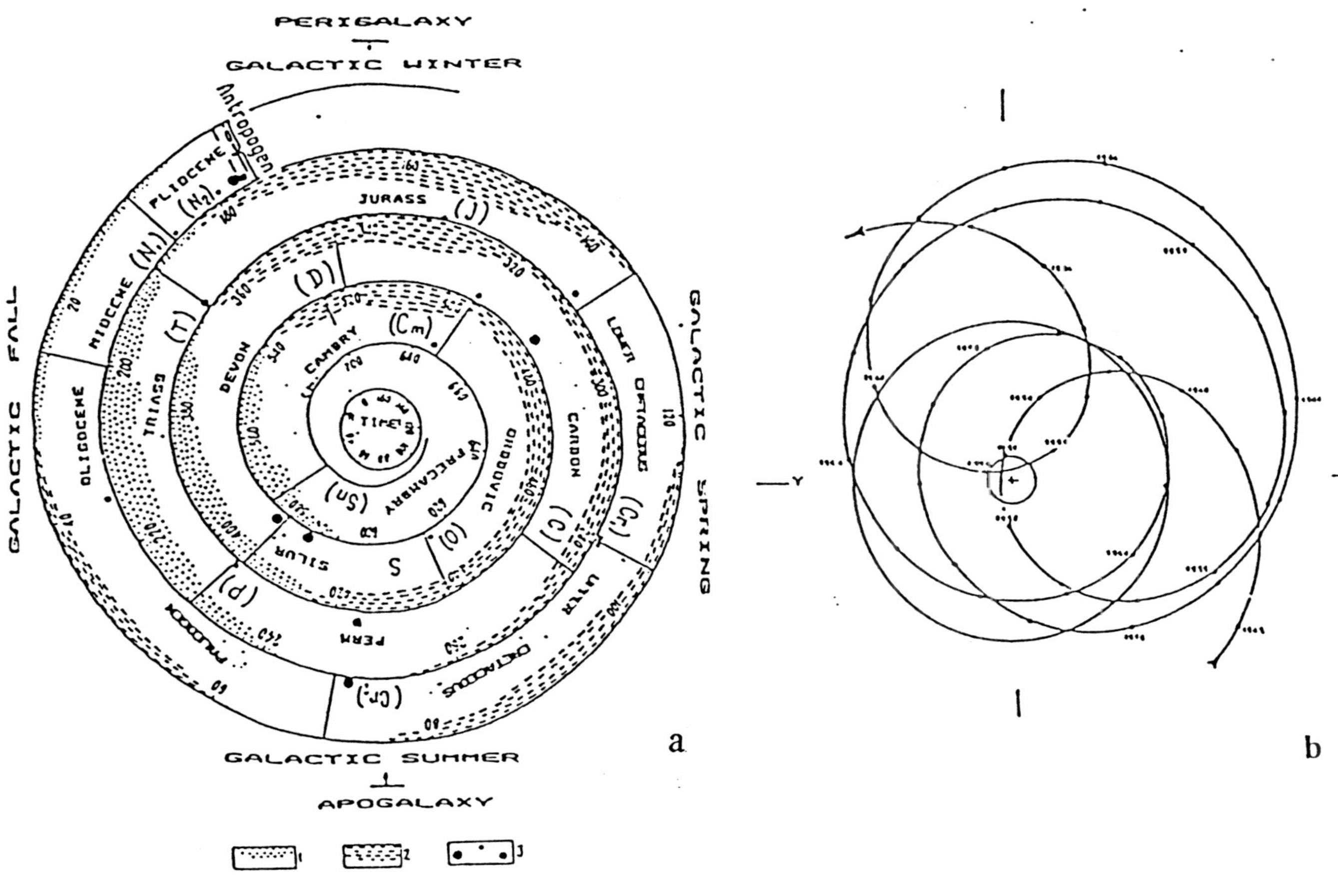

Fig. 12 *Abnormal periods of the solar system movements, and the seasons of galactic year on the Earth during last 660 million years. 1. The main geocratical epochs, 2. The main talassocratical epochs. 3.The largest phases of the tectogenesis. (after Tamrazyan) (a), Master cycle of solar system for the years 1945 to 1995. Small circles indicate the position of the center of mass of the planetary system which sometimes is above or below the sun's surface. (after Landscheidt, 1995) (b)*

12b the master cycle of the solar system is shown. These all testify to, the structural self-hierarchy of the celestial bodies.

So, the oscillation of the galaxy center's orbital angular momentum around mean, during the galactic years' cycle, can form constellations' appropriate long length's cycles, which in turn generate their primary, secondary, and tertiary bodies short-length cycles until the accomplishment of the long -length cycle. Such structural and functional organization is stipulated by the energy's optimum distribution (Sarafyan 1993,1995). Indeed, in order to restore the balance after every excitement, the magnitude of the galaxy center's orbital angular momentum during the cycle should be distributed, liquidated and dissipated. The distribution takes place between two components: the galaxy center's rotational angular momentum's, and constellations' primary bodies' orbital angular momentum's. Transformed to the primary body the movement's quantity is distributed between two components again: the primary body's rotational angular momentum's, and the secondary bodies' orbital angular momentum' and so on. Otherwise, we have hydrodynamic type's dissipation: large scales constantly generate small scales. The only difference here is in the definitively and steadiness of scales by which take place dissipation. This means that the original magnitude of the movement's quantity (and any other endogenous influence) gradually is transformed, dissipated, and the balance is established quickly by them: the system regularly becomes ready to react and transform the energy's succeeding portions. At this point the primary and decisive force is external influence to which the system permanently is adapted maintaining existing energy's distribution and orderliness. Some balance equations of such relations will be analyzed below.

4.Geophysical Alterations' Direction

The mass of the Earth, its composition, density, diameter, speed of rotation and circulation, the existence of a single satellite and other characteristics are stipulated, not by themselves but by specific conditions of the space. They are stipulated by other planets, and mainly, by moments of revolution and rotation which are compelled by the Sun. Similarly, the generation of dividing borders inside the Earth, are stipulated neither by alterations of the chemical composition of various depths of the Earth nor by phase transitions of its substance, (Khain,1964; Odinstov,1980) nor by gravity differentiation. They are caused by external conditions of the space; by their demands, requirements and developments. Alterations of the chemical composition in different depths and the physical-mechanical properties of the shells, gravity differentiation, distribution of the $\mathbf{p}$, ρ, η, μ, $\mathbf{g}$ and other parameters are the result, but not the cause of the established equilibrium in the conditions of space, just as minerals create rocks in accordance with physical-chemical conditions of the surroundings. Rules and causes of the ore formation are not in the ore minerals or are bodies, but in the conditions of the medium that inevitably generate accumulations of the ore. It is possible to continue these thoughts in another direction, but we will restrict ourselves to these. Thus, the Earth's structure, its shells, and its other features are a direct result of the arrangement and adaptation to the mutual positions of the celestial bodies. They are also consequences of the conditions of the solar system and depend on the development of these conditions. This conclusion is not new. It springs from general physical consideration. Its feature is that it is derived exclusively from the analysis: (1) the pumping of energy and forces, their transformation and dissipation in the shells consecutively, and (2) the structural features of the space and its permanently acting and transmitting angular moment and impulse.

5. Shells' Functions

The question is how the gravity interaction, the transfer of angular momentum, its pumping, transformation and dissipation in different shells and between them take place. Without general balance equations and mathematical simulations it is extremely difficult to understand what occurs in the Earth's depths. How are the above mentioned interactions realized? What are the forms and amounts of the pumping and transformation of energy, as well as the order of geophysical events? In order to reproduce the behavior relative to the various relaxation's of embedding shells, and to write the equations of dynamics for them, it is necessary to have a complete notion about the structure of space. It is necessary more specifically, to understand the spectrum of influences that is significant for the Earth and its shell's functions. However, a rational interpretation of the variety of facts about the numbers and dimensions of the shells, as they relate to the cycles of the solar system has still not been found. While the existing data on the physical-mechanical properties and material composition of the shells are considered satisfactory (Gutenberg,1959; Magnistky,1965; Bott,1982) the same cannot be said about the shell's functions. Indeed, from the geophysical point of view, the shell's function is not quite clear. Their character, in the division of the planetary labor along geological time has not been explored. The answer here is by no means in the geological data, in the peculiarity of the shells or in their developments; but rather in the realization of the permanent active energy and mass transfer processes in the liquid shells and of the rheologic, geodynamic regular motions and dislocations in the solid shells, which pass cyclically and do certain work. In the simulation, the main objective is to take into account the gravity coercion, its mechanism of transformation and dissipation in the Earth's shells and crust. In the case of gravity waves the Earth represents an open system: they penetrate through the planet and by means of the viscous currents and friction, are transformed and dissipated, having been converted into thermal energy. On the other hand, as regards physical substance Earth is more or less a closed system. The closeness is a consequence of the above mentioned functions of transformation and dissipation for the shells. Actually, a substance of a certain volume is permanently subjected to convection, advection (magmatism), phase conversion, differentiation and other energy transfer processes (in the case of a liquid state). Or it is decomposed to the mineral phases and ions, is periodically compressed, consolidated and melted into solid (in the case of solid state), (Hlobustov, 1971; Odinstov, 1980).

6. Simulation

Let **R** be the quantity of energy, received by the core as the result of gravity interactions. Since the core's relaxation time is $\tau = \mathbf{10^7}$ years, the period **T** of interaction should be more than $\mathbf{10^7}$ years. Assume that, as a result of convection, part of the total amount of energy entering the core is distributed as follows: part $\mathbf{k_1}$ is spent to increase its mass: part $\mathbf{m_1}$ is spent to maintain the core's structural and functional properties, and part $\mathbf{c_1}$ through interaction is transmitted to the mantle. The core is passive and barely re acts to excitements of short duration. Entering the mantle, portions of energy (independent if they come from core or external medium) are processed in the same manner. Namely, part $\mathbf{k_2}$ is spent to increase the mantle's mass; part $\mathbf{m_2}$ to protect the mantle's structural and functional properties; and part $\mathbf{c_2}$, through interactions (convection of the second order), is transmitted and spread to the lithosphere and crust. Basically the same occurs with the energy entering the lithosphere. Taking into account that $\tau = \mathbf{10^6}$ years for the mantle, it can be assumed that energy flow into the mantle occurs every $\mathbf{T>10^7}$ years. At the same time the mantle is passive to influences of short duration and barely reacts to them. Sub-harmonic resonance is not examined due to their unstable influence.

From the geological perspective, the most interesting issue is the energy transformation in the lithosphere. While in the internal shells the energy transformation takes place primarily by convection and turbulence. In the lithosphere, due to its hardness transformation generates rheologic, geodynamic regular movements of different scale. It is not clear whether they are the motions of hydrodynamic character or not (scales of large dimensions generate scales of small dimensions). Whether they are accompanied everywhere by magmatism (advection), folding or other processes.

To determine the case, it is necessary to verify and classify structures of the geological cycles (Sarafyan, 1990,1984), especially their relaxation phases. Usually processes of restoration at the starting point state are compound and multistage. For instance, in the active media certain processes arise following excitements that constantly replace each other at the time (Burkov,1974). Restoration processes in the lithosphere are very involved and complicated. Their exact definitions and classifications do not even exist. Therefore, it is difficult to say whether we should consider the slow rising, folding, magmatism, earthquakes, sedimentation and other geological phenomena as components of a unified restoration process. Nevertheless, the identity, the definite parallelism and the repeated appearance of the above-mentioned geological processes allow one to accept such a possibility.

Let us continue to describe the mathematical model. During the gravity coercion, which has a periodicity of more than 10^6 years, the core gets a definite portion of energy that is processed, pumped consequently through the shells, transformed, and dissipated. The same happens with coercion in the other shells. Coercion with period $\mathbf{T} = 10^7$ years represents the upper limit for interactions of Earth within the surrounding medium. For simplicity's sake neither the lower excitements nor the excitements of super-low frequencies is considered here. It is very important to determine the values of the coefficients $\mathbf{k_i}$, $\mathbf{m_i}$, $\mathbf{c_i}$. From the physical considerations it follows that $\mathbf{0 < k_i, m_i, c_i < 1}$. In atmospheric or ecological system 70-80% (and sometime 95%) of the energy entering the system is usually dissipated, having been converted into thermal energy (Lorenz,1964; Odinstov, 1980; Rossby, 1951; Sverejew,1978). These facts may be useful in the evaluation of coefficient $\mathbf{m_i}$. About 75% of the energy entering the shells can be attributed to that coefficient. At the present time it is difficult to say how to estimate the coefficients $\mathbf{k_i}$ and $\mathbf{c_i}$. The condition $\mathbf{0 < k_i < 1}$ is important for another consideration too. It permits confirmation of Dirac's assumption about the permanent growth of material inside the Earth (Dirac,1972,1973). Based on the above reasoning, the general balance relation of energy and material for the three-level system, will have the following form

$$\frac{dU}{dT} = R - m_1U - k_1 U - c_1UV ;$$

$$\frac{dV}{dt} = k_2UV - m_2V - c_2VW + f_1 ;$$

$$\frac{dW}{d\tau} = k_3 VW - m_3 W + f_2 \qquad (3)$$

where $\mathbf{T > t > \tau}$, and where they represent predominant periods in the 1, 2 and 3, shells, respectively. $\mathbf{f_i}$ is the exogenous excitement. The physical meaning of the condition $\mathbf{T > t > \tau}$ is that the processes pass through the shells at different rates In order to analyze the system (Eq.3) one should change the independent variables and find more precise values for coefficients initial and limited conditions. How ever, we will consider the generalization of the advanced system (Eq.3) instead. Let us assume that, in the shells, diffusion processes are carried out which have velocities higher than the velocities of excitement. Then, system Eq.3 is converted to the following system equation

$$\frac{\partial U}{\partial T} = D_u \Delta U + f_1(U,V) ;$$

$$\frac{\partial V}{\partial t} = D_v \Delta V + f_2(V,W) ;$$

$$\frac{\partial \mathbf{W}}{\partial \tau} = \mathbf{D}_w \, \Delta \mathbf{W} + \mathbf{f}_3(\mathbf{W}) \qquad (4)$$

where $\mathbf{D}_u$, $\mathbf{D}_v$, $\mathbf{D}_w$ are the diffusion coefficients in the corresponding shells, Δ is the Laplace's operator and, $\mathbf{f}_i$ is the functions of connection and interaction between **U**, **V** and **W**. The system Eq.4 in many respects is more general and describes phenomena of a wider class than Eq.3. For instance, Eq.4 is very important in the analyses of morphogenesis and some other problems of modern geology.

The system of equations will basically be the same, if the type of motions will be different: such as convection, advection, whirlwind etc. The origin of similar movements and energy-exchanging processes is natural. The origin of them should be substantiated. For instance, the averaging and balance with diffusion requires a large amount of time for a medium of large dimensions. Hence, it is not effective for that purpose. In such cases, another balancing form, such as, convection, are necessary. Obviously, diffusion again will be converted to convection if the intensity of the energy becomes critical for the medium. The averaging or balance by the convection (more so, by the diffusion) obviously does not affect media, like hydrosphere and atmosphere, where there are periodical accumulations of energy (exertion, strain, effort) in the separate places. In such cases more rapid processes, like vortex currents are necessary. Vortex currents (cyclones and anticyclones) quickly move accumulated strain long distance from their rising places (centers) and restore equilibrium (Lorenz, 1964). Finally, the averaging (balance) consumes most of the rapid processes, like turbulent currents or movements for the medium (even, homogeneous) where the magnitude of the relaxation is bigger then the length between impulses (Hydrodynamic Instability, 1981; Hayashi,1966). So, relaxation, dimensions, distant between excitements and other parameters are stipulated the types of the motions. They have important value and significance, in the division of the planetary effort (labor). Their alterations make it necessary to adjust to different functions or their alterations. The most diverse functions must be adjusted. Thus, the appearance of convection, advection, vortex currents, turbulence and other forms in the energy pumping do not change the basic concept; systems constantly must realize dissipation, conversion and transmission energy (substance, signal) at the interactions.

7. Conclusion

System Eq.4 is more general than Eq.3, but it is not adequate to reproduce the interactions of the above system. It describes only the excitement's movement (its front) and does not include its reconstruction processes, which are constantly rising behind the excitement front (Sarafyan,1984). The laws of excitation movement are not as important as those generated by process restoration, its structures, its displayed forms and cyclicity for geophysics, and especially for geology. Further, in Eq.4 no reason is given for the source of diffusion processing. Finally, the existence of convection in the core and mantle already means that convection presents itself as auto-oscillation. Otherwise, it is a relaxation process and consequently "homogenous" equations of the system Eq.3 do not reflect reality. These are necessary other modification of the Nave-Stokes' equations. Nevertheless, system equations like Eq 3 and Eq.4 are useful. First, due to the hierarchy, they unite geophysical different range processes, interpret them from a unified point of view, and indicate a general trend. They provide also an orientation for the geophysical alterations.

Second, system Eq.3 or Eq.4 provides some confirmation of Dirac's hypothesis on rising material in the core (and, seemingly, not only in the core), at any rate for a certain period of development. Such phenomena are widely observed and are considered natural in ecological and biological systems (Volterra, 1935; Sverejev,1978). Earth is a developing system, because its rotary speed decreases, whereas crust's heterogeneity, power and hardness, the tectonic differentiation increase. So, it should display a similar tendency; to receive (by a frequency), transform and transmit (by another frequency) energy, substance at the interactions. That is to say, Earth is an ordinary running (flowing) system, with all its sequences. At last, perhaps, granites' problem is a consequence of Dirac's hypothesis.

Third, it allows one to judge a burning question of modern physics. The problem is in the existence and possible spectrum of gravity waves. Indeed, what represents the external medium of the Earth, in which it has existed for 5 billion years and with which it interacts? From the stability relation of the system (Earth-space) it follows that there are equilibrium interactions at the geological axis of time, between the Earth (and, hence its shells) and the corresponding external excitements. According to this discourse pumping, alteration of momentum and dissipation of the external influence constantly take place in the shells. They create the cyclicity of the long range tectonic movements, magmatism, earthquakes, gas and oil formations, sedimentation and geological other processes as well as the evolution of the chemical, mechanical and physical compositions of the shells. Therefore, from the self-hierarchy shell's parameters (measures, relaxation and others) follow the self-hierarchy of the relations of the external influences and the excitements, which are essential to the Earth. This conclusion must be regarded as an attempt to determine the truth. Usually, a medium's conditions determine the functional and structural features of the system, but not vice versa. So, only some ranges' of excitations for the Earth have planetary value. If the sequence of the numbers of years: $\mathbf{10^7}$, $\mathbf{10^6}$, $\mathbf{10^5}$, $\mathbf{10^4}$, $\mathbf{10^3}$ (or another sequence of numbers), represents the magnitude of the relaxation of the earth structures (whole planet, core, mantle, crust and other spheres) then there exist external particular sources of energy, with respect to the above structures. The changes by any above periodicity, inevitability entails (requires) alteration (reaction) of those spheres or structures, which by theirs functions are balanced previously change. What relationship exists between excitations of the indicated ranges and gravity waves? Perhaps, examination of that question will help one to understand more about the nature of gravity waves, their features and possible spectrum, which apparently depends on the radii and compositions of celestial bodies.

3.3. SIMPLE SEDIMENT CYCLES

In the preceding sections it was mentioned that calm or restoration stages of different-scale tectonic cycles are heterogeneous and multi-nature processes in the geological sense. Restoration stages of tectonic cycles are necessarily accompanied by the following geological phenomena: sedimentation, folding, magmatism (with effusive and intrusive functions). According to some investigations each of the mentioned processes is, in its turn, of multi-stage, cascade and subordinated processes.

Among the studies of cyclic geological changes the study of sedimentation and stratification cycles occupies a special place. Good preservation and wide spreading of stratification process results comparing to other geological phenomena's traces make them reliable indications for studies of other geological cycles. Composition, nature, types (symmetric, non-symmetric) of sedimentation cycles are studied and processed comprehensively and in details. Here only quantitative questions of their structures and formations are of interest. These questions will be discussed basing on the data of so called flinch (non symmetric) sediment formations which are studied best of all in the field of sedimentation. The main source for us in this discussion will be (Sadoyan, 1983).

Two lines are now observed in study of sedimentation problems by statistical methods. In the first line hidden periods are found out in sediments by means of well-known mathematical methods (expansion into Fourier series, spectral expansion), revealed periodicity and curves are grounded within the frames of existing geological facts and their geological contents and meanings are revised. The attention in the works of the second line is focused on statistical characteristics of layers and formations: distribution of their main components X, E, D, A_x, $\tilde{A}$ and their double and multiple correlation $\tilde{A}_{xy}$, $\tilde{A}_{x,y,z}$,... The literature concerning sedimentation cycles allows to estimate conditions of layers formation in different sections, their nature and time (limits), and in the simplest cases hidden periodicity (Anarin and others, 1984), to establish relations between both separate layers and dynamics of plate horizontal motions and accumulations (Korogodin and others, 1984). Quantitative studies of material-structural composition of layers, non-stationary and directness of accumulation process are of great importance. For these purposes the Markovian processes and other branches of mathematics are widely used (Sarafyan,1984, Harbrook, 1974). Below, such a discussion is to be carried out basing on the data about Paleocene sediments of flesh formations of Armenia, trying to use, in the course of this discussion, matrices of Markovian processes, or rather, layer alternation or transition probabilities in order to distinguish cycles of different orders in them.

Flysch formations under discussion are subdivided into three sub-formations from which 28 characteristic sections are selected. The geological description's material compositions, thickness and other characteristics are detailed in (Sadoyan, 1983). We are interested in alternations of different layers and groups of events in suite succession of the mentioned sub-formations, and regularities of their distribution. For this purpose the tables of layers and suite's recurrence frequencies are made up (for each section and the matrices of transition probabilities are determined. Six typical matrices from these 28 matrices are chosen for discussion. Their conditional numbers are 14, 6, 20 and 7, 28, 24. The division of the matrices into two groups is related to containing and non-containing of a large number of 0 and 1 in them. The matrices are of the following forms:

$$P_{14} = \begin{matrix} 0 & 1 & 0 & 0 \\ 0 & 0 & 1 & 0 \\ 0 & 0 & 0 & 1 \\ 0.35 & 0.5 & 0.15 & 0 \end{matrix} \qquad P_6 = \begin{matrix} 0 & 1 & 0 & 0 \\ 0 & 0 & 1 & 0 \\ 0 & 0 & 0 & 1 \\ 0.05 & 0.15 & 0.80 & 0 \end{matrix}$$

$$P_{20} = \begin{matrix} 0 & 0 & 0 & 0 & 0 \\ 0 & 0 & 1 & 0 & 0 \\ 0 & 0 & 0 & 1 & 0 \\ 0 & 0 & 0 & 0 & 1 \\ 0.15 & 0.14 & 0.17 & 0.54 & 0 \end{matrix} \qquad P_7 = \begin{matrix} 0 & 1 & 0 & 0 \\ 0 & 0 & 0.97 & 0.03 \\ 0.15 & 0.61 & 0 & 0.24 \\ 0.21 & 0.79 & 0 & 0 \end{matrix}$$

$$P_{28} \quad \begin{matrix} 0 & 1 & 0 & 0 & 0 \\ 0 & 0 & 0.93 & 0.07 & 0 \\ 0.1 & 0.15 & 0 & 0.75 & 0 \\ 0 & 0 & 0.025 & 0 & 0.975 \\ 0.5 & 0.19 & 0.03 & 0.28 & 0 \end{matrix} \qquad P_{24} = \begin{matrix} 0 & 0.9 & 0.1 & 0 & 0 & 0 \\ 0.1 & 0 & 0.99 & 0 & 0 & 0 \\ 0 & 0 & 0 & 0.4 & 0.28 & 0.32 \\ 0 & 0 & 0 & 0 & 0.45 & 0.55 \\ 0 & 0.24 & 0.17 & 0.12 & 0 & 0.52 \\ 0.28 & 0.28 & 0.24 & 0.08 & 0.12 & 0 \end{matrix}$$

In the sub-formation I and II the basic layer forming elements are: 1) sand; 2) siltstone; 3) marlstone and 4) clay; in the sub-formation III coarse sands are added to them. The geological meaning of the transition probability matrices is as follow: in the structures of group **I** layer of sand is followed by siltstone layer with probability 1, then followed by marlstone layer with probability 1, then the marls-tone layer is mandatory followed by clay. Then clay layer is retransit to sand layer with probability 0.35, to silstone layer with probability 0.5, and finally, to marlstone layer with probability 0.15. The similar pattern is also observed in the section 6 and 20; the transitions to silstone from sand, to marlstone from silatone and to clay from marlstone are mandatory. The next group of sections in the matrices 7, 28 and 24 such direct transitions do not exist. Here one layer is changing to another one with a certain probability that is lower than unity. As this takes place such transitions are observed during each one-step transfer.

In the geological sense the most important are probabilities of transitions after n steps. In this way so called regular (final) probabilities of layers in suite alternations which are specified by the fact that a transition from any state to another becomes independent on initial state, i.e. all lines of the matrix become the same, are determined. Probabilities of multi-step transition are determined by raising the initial transition probability matrix to a power. In the process the required power is obtained when the initial matrix is not changed else during transformations, that is the farther raising it to a power does not change the matrix.

The regular matrices for the above mentioned ones are the followings:

$$P_{14}^{11} = \begin{matrix} 0.13 & 0.32 & 0.30 & 0.26 \\ 0.10 & 0.27 & 0.36 & 0.30 \\ 0.10 & 0.24 & 0.31 & 0.36 \\ 0.13 & 0.28 & 0.30 & 0.31 \end{matrix} \qquad P_{7}^{6} = \begin{matrix} 0.8 & 0.41 & 0.41 & 0.10 \\ 0.8 & 0.41 & 0.41 & 0.10 \\ 0.8 & 0.41 & 0.41 & 0.10 \\ 0.8 & 0.42 & 0.41 & 0.10 \end{matrix}$$

$$P_{6}^{5} = \begin{matrix} 0.04 & 0.17 & 0.15 & 0.64 \\ 0.03 & 0.14 & 0.17 & 0.66 \\ 0.03 & 0.13 & 0.14 & 0.70 \end{matrix}$$

0.04 0.14 0.14 0.70

$P_{28}^{11}=$

0.09	0.24	0.23	0.16	0.29	0.05	0.10	0.17	0.32
0.17	0.18	0.24	0.26	0.16	0.05	0.10	0.16	0.36
0.10	0.23	0.18	0.23	0.25	0.05	0.10	0.16	0.33
0.15	0.17	0.26	0.21	0.23	0.05	0.10	0.16	0.35
0.14	0.23	0.18	0.25	0.20	0.05	0.10	0.16	0.34

$= P_{20}^{4}$

$P_{24}^{7} =$

4.8	0.18	0.27	0.16	0.18	0.26
0.9	0.19	0.27	0.15	0.18	0.27
0.9	0.19	0.26	0.14	0.16	0.27
0.9	0.19	0.27	0.14	0.16	0.24
0.9	0.19	0.27	0.15	0.16	0.25
0.9	0.18	0.28	0.16	0.17	0.25

Some matrices do not acquire their final values because of deficit of computer calculation time, the lines of given matrices characterize balanced shares of different states after 5 to 11 steps. These lines represent observed probabilities P_1, P_2,...,P_n of states a_1,a_2,...,a_n independent on the choice of an initial state. Thus, for the section 24 the initial state is: 1) coarse sand, 2) fine sand, 3) coarse siltstone, 4) fine siltstone, 5) marlstone and 6) clay. The observed probabilities are equal to 0.88, 0.18, 0.28, 0.25, 0.18 and 0.25 at the number of transition steps equal to 7. Hence, the 7th power of the initial P_{24} matrix is represented by P_{24}^{7}.

Section transition probability's matrix may be used to find out cycles of different frequencies. In this way for any section a certain number of cycles, suites with definite observed probabilities are determined formally. Such cycles for the mentioned sections 14,6 and 20 are given below:

P_1=0.5	1-2-3-4-2	q_1=0.80	1-2-3-4-3
P_2=0.35	1-2-3-4-1	q_2=0.15	1-2-3-4-2
P_3=0.15	1-2-3-4-3	q_3=0.05	1-2-3-4-
P=1.00		q=1.00	

$r_1=0.54$ 1-2-3-4-5-4

$r_2=0.17$ 1-2-3-4-5-3

$r_3=0.15$ 1-2-3-4-5-1

$r_4=0.14$ 1-2-3-4-5-2

———

$r = 1.00$

The cycles with their corresponding probabilities for the sections (matrices) are given in the following form:

$P_1=0.5917$	1-2-3-2	$q_1= 0.3619$	1-2-3-4-1	$r_1= 0.0792$	1-2-3-6-1
$P_2=0.1839$	1-2-3-4	$q_2= 0.2026$	1-2-3-4-5-4	$r_2= 0.0548$	1-2-3-4-6-1
$P_3=0.1455$	1-2-3-4	$q_3= 0.1485$	1-2-3-2	$r_3= 0.0284$	1-2-3-5-6-1
$P_4= 0.0488$	1-2-3-4-5	$q_4= 0.1375$	1-2-3-4-5-2	$r_4= 0.0348$	1-2-3-4-5-6-2
$P_5=0.0237$	1-2-4-3	$q_5= 0.0287$	1-2-3-4-5-3		
$P_6=0.0061$	1-2-4-1	$q_6= 0.0186$	1-2-3-4-3		
———		$q_7= 0.0099$	1-2-3-1	$r = 0.2072$	
$P=0.9997$		———			
		$q=0.9007$			

The number of all possible cycles for the matrix 24 is equal to 27. But here only those cycles are given, for which the probability of existence is higher than the 28 average value, i.e. higher than r=0.037. The remaining cycles for the matrix are not given here because of their low probabilities.

Usually cycle layer alternations of different orders in a given initial layer succession are distinguished by means of computation of occurrence frequencies of **I, II, III** and other component elements. This principle was also applied in the presented calculations with the only difference that usually cycle order is determined directly through the table in parallel to motion, but here-by means of determination of the products of all possible transition probabilities, taking by one element from every line and every column. The presented scheme is convenient because it allows to calculate possible cycle separately and to classify their observed probabilities of each cycle in the total volume of oscillations and in the complete field of probabilities. In order to determine energetic contributions it will be necessary to take account of layer thickness, distances of supplying sources, rate of accumulation and other characteristics. It is easy to be assured in efficiency of the presented method by means of considering the data for the sections 14, 6 and 20. Various cycles and frequencies determined by the presented method are given below in the table 7:

Table 7.

Section	Order of cycle				Total number of cycles
	I	II	III	IV	
	-	8 p=0.32	12 p=0.48	5 p=0.20	25
	-	3 p=0.14	11 p=0.15	54 p=0.80	67
	-	6 p=0.16	5 p=0.13	27 p=0.71	38

The calculations made by both methods for the sections 14 and 6 are the same. But in the section 20 one suite which is recorded by the presented method suite not distinguished by the usual method. The difference between these two methods is already essential when considering the sections 7, 28 and 24, i. e. when matrices contain a small number of 1 and 0. The calculations of cycles by the method have not their prototypes or similar ones in the groups of cycles calculated by the usual method. Thus, the usual method determines 2, 3 and 4 cycles of different orders for the sections 7, 28 and 24, correspondingly, while the proposed method calculates their numbers equal to 6, 10 and 27. The reason is that usually the first, second and third component elements are only taken into account when calculating cycles of different orders, their existence or absence is the basis for calculations. The last elements of cycle, the composition and duration aren't paid a special attention, which results in unification of some cycles-suites into the one united cycle and represent one unit. Thus, the following cycles 1-2-3-1, 1-2-3-4-1, with probabilities p_1=0.1455, p_2=0.1408, p_8=0.0061, distinguished by the presented method for the section 7, being united into one cycle, will result in the observed cycle with probability p=0.2004. The cycle of **III** order, calculated by the usual method, which has probability of observation p=0.2, is corresponding to this cycle. The unification of suites 2-3-2, 2-3-4-2, 2-4-2 of the same section with probabilities of observation p_1=0.5917, p_2=0.1939, p_3=0.027 results in one united cycle with observation probability p=0.8093. This cycle is corresponding to the cycle of **IV** order calculated by the usual method, the probability of its observation being p=0.79.

Thus, determination of cycles by the proposed method is more general and convenient and there is not, basically, differences between the results. Some deviations of the result are consequences of the fact that composition, duration and last element of cycles are not taken into account when calculating them by the usual method.

What is the relation between an order of cycle and its observation probability? As the most of the presented examples show, the observation probability of cycle increases with the growth of cycle order. This result is not surprising, but it actually confirms the thought that cycles with short duration should have higher observation probability and be involved in cycles of long periods (duration) as components.

Let us now consider possible relations between sedimentation cycles and forces giving rise to them. Usually improvements of geochronologic scales, revision and distinguishing of hidden periodicity limit is performed without taking account of specific features of forces and mechanisms giving rise to them. Neglecting of these forces with respect to layer forming cycles and their achievements cause some distrust of other geologists. And this occurs in the case when cycles' problem is very important one from the viewpoint of dynamic systems, because every time they transfer and transport the energy of external excitations and dissipate it providing a stability of observed scales in the conditions of changing medium. The equilibrium established at the expense of optimum distribution and compensation of excitation energy forms new space-and-time periodicity and structures. For this reason in order to introduce and apply geological cycles studies in other branches of geology, it is necessary to connect the problems of origin and structure of cycles to their tectonic, geological, earthquake, transgression and regression changes. The questions arising here need comprehensive investigations. The difficulty is that the nature and characteristics of forces giving rise to space-and-time periods have subordinated structure and represent cycles by themselves. This property is common for all (long period and high frequency) conditions and forces of sedimentation. And even more, it is true not only for sedimentation but for cycles of other nature too. Thus, the concept of multi-cycle recurrence involves some elements of arbitrariness, if they are studied without consideration of forces giving rise to them, and more precisely, structures of energetic expenses.

According to the accepted model of flish sediment's formation, alluvia in ocean shelves can not be accumulated unlimitedly. Reaching a critical mass, from time to time they loss their equilibrium or slipped due to underwater shocks, and rushing down spread in deeper and more central areas of the ocean bottom, covering everywhere flat-bottom valleys. These alternations accumulations and landslips of river alluvia and substances are not identical on the geological time axis because forces giving them rise are of multi-nature and different periods, therefore arrangement of layer groups are different and not identical. The laws of layer alternation, beyond doubts, may somewhat elucidate spatial-and-temporal distribution of forces of origin. Particularly, layer groups and cycle order may be useful when restoring the regularities of given sand's ***palaeoseismisity.***

Small (higher order) cycles will correspond to low-magnitude shocks, and large (lower order) cycles to high-magnitude ones. Since thickness of sediments is equivalent to time intervals of river brought substance accumulation then basing on their analysis it is possible to reconstruct periods and magnitudes of underwater palaeoshocks. In the case of explicitly of these relations it is possible to estimate palaeoelastic stress distribution, order of their accumulation and dissipation. If such a mechanism is true, then these considerations will allow determine the lower (possible smallest) limit of palaeostress of a region under investigation, to estimate plastic and elastic properties. In general, in the case of the same conditions of sedimentation it is possible to judge, basing on sediments, on their supplying source (focus) characteristics, particularly, on its nature distance, altitude, washing out rates, etc. Thus it could help to construct now yet weak but stable connection between different branches of geology.

CHAPTER IV

WORLD OCEANIC CYCLES

Cyclic changes of the ocean are caused by interactions with adjacent media. At the first place among them should be mentioned the Earth's crust, the atmosphere, gravitation of the sun and other heavenly bodies, interactions of tidal, thermal and other natures. It seems that contributions of the Earth's crust oscillations, magmatic contributions and other cyclic changes into forming of the ocean behaviour are small. But it is not so. In the ocean bottom in the zones of the Earth's crust expansion, i.e. rift areas, thermal flow is higher by a factor of 6 to 7 than the average value of this flow calculated for the whole planet. In the Red Sea the temperature of rift near-bottom waters reaches up to 60^0C.The ridges, where underwater volcanoes are active, are distinguished in the ocean bottom. No less important is underwater periodic shock, earthquakes, which is one of the plausible factors of internal wave's origin. Then, the magical triangles of the Earth's crust, being the most tectonic active sites, where thermal, elastic and other discharges of stress are more intensive, directly transfer changes through oceanic near-bottom layers causing new distributions of hydrophysic parameters. Geophysical properties in these sites are drastically differing from those in adjacent areas. For example there is a particular positive abnormality of the gravitation field up to 200 mgal in the Bermuda triangle. The abnormality in magnetic field is much greater and amounts up to 1.000 to 1.500 mgal, in separate regions up to 5.000 mgal. Here, the ocean surface is depressed by 25m, than is the mean level of the ocean, etc. The concentration of mineral salts in near-bottom waters of the Red Sea is higher by a factor of 50 thousand than for ordinary sea's water. This is the reason why water does not rise to the surface despite its high temperature. Sometimes in rift zones water is heated up to 200^0C and periodically ejected under a pressure of approximately 300 atm (Gavrilov 1978). The rotation of the Earth around its axis gives rise to the oceanic changes of the same period, which affecting the rotation through a feed back cause its non-uniformity. Just in the same way the effects of the atmosphere, the Sun and other heavenly bodies do not remain without consequences. In these cases other regular motions and changes occur, which now neutralise and balance these excitations.

Thus, the ocean receiving and absorbing external multi-nature excitations (material or energetic) constantly performs a certain work: it reconstructs, redistributes and reforms its morphological structure according to a some law, trending to dissipate in these ways effects of external excitations. Therefore the internal structure in the ocean activity isn't a luxury, but a necessity for realisation of excitation neutralisation and restoration of violated equilibrium within the limits of observed scale in the conditions of changing external medium.

1. Oceanic Cycles' Nature

What are oceanic cycles? They are oscillations of its surface level (local, regional and general), large-scale rotations and axes, expenditures and velocities of ocean currents and regular changes of linear (fronts, boundaries, layers) and non linear (vortices, rotations) formations, galactic boundaries and thermal, density and salinity (local, regional and general) distributions. These changes are characterised by short and nonregular, daily and long-term and secular components, which result in infinite variety of interactions up to random phenomena and force to think that direction and unity of structure and internal interaction can not exist in this variety. But oceanic currents and rotations and other morphological structures represent complicated systems of self-organization. Hence, cyclic changes of their behaviour are strictly determined and depending on external (in respond to observed scales) excitations.

2. Oceanic Cycles' Structure

Problems of oceanic cycles' structures are at much more developed level than the similar problems of solar or geological cycles. All cyclic changes of the Ocean, beginning from the smallest ones, i.e. of micro-and meso-scales up to the largest, i. e. tens or hundred of years, represent one united system. The main drawback of the concept existing earlier is that frequencies, amplitudes and initial phases of all observed changes were considered as completely independent on time and each other. The first approximation of such an approach to these changes resulted in their infinite diversity, including even random events, and their non-determinity, which plagued to see organiza-tion, and unity of internal structure and interactions (motions), which are displayed by a repetition of a series of changes caused by same excitement. In 1960s this concept was replaced by a new one, particularly including the principles of comprehensive modulation which, allowed to study time-dependent changes of period, amplitude and phase, taking account of motions and dependence of changes on each other. Then it was found that the distributions of energy-consuming oceanic cyclic changes scales aren't random ones, but are enclosed one another in a specific way, let us say, by the automodal principle. Changes of given scale are formed in interactions with specifically close next scale. In other words, oceanic cycles, similarly to biological rhythms, are provided with a certain organiza-tion, line of changes and play a certain role in meteorological sharing of works. It is displayed by the fact that the phenomena of observed scale are constantly occurred and developed on the background of larger-scale phenomena and dissipating of transferred energy. Thus, seasonal changes of known water velocity field, temperature and salinity values are observed on the background of these values, and slower running processes such as tides and internal waves, formation of the ocean morphological structures and so on are observed on the background the former ones.

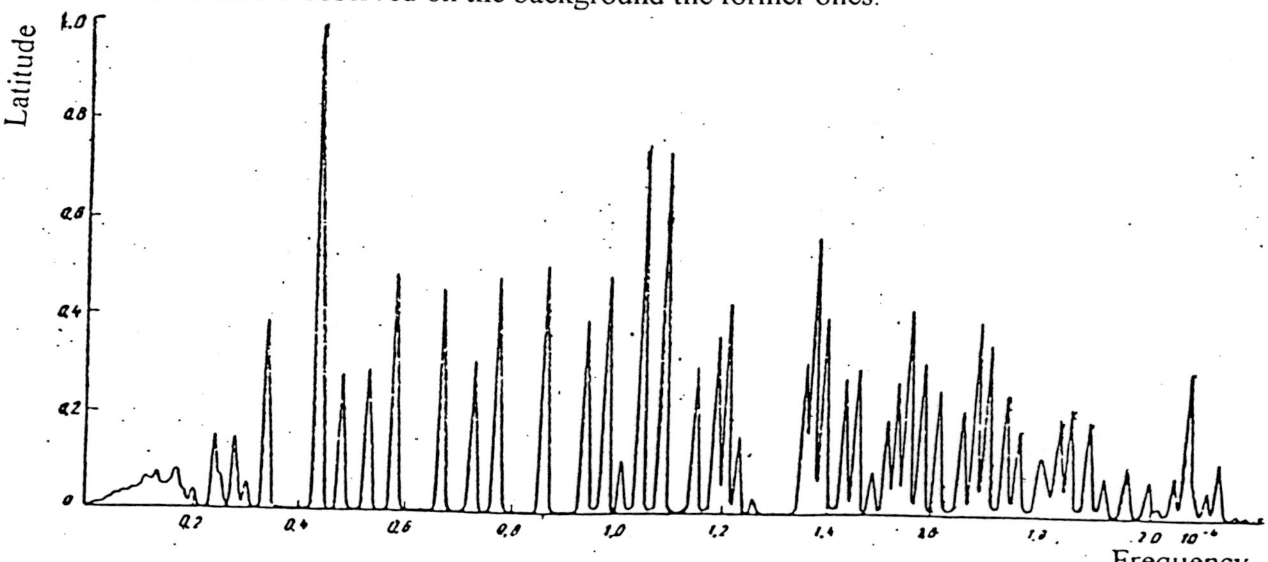

Fig. 13 *The specter of the World Ocean high frequency oscillations. The value of circular Frequency is equal $0.7*10^{-6}$ sec^{-1} which ≈ 24.6 hours and $1.4*10^{-6}$ sec^{-1} ≈12.3 hours*

The considered motions and turnovers are determined and the best of all the possible motions (optimum motion) in the sense that it transfers a substantial share of substance (energy) to the required site in the required time, restores the equilibrium violated during a cycle within the observed scale limits. In other words, the well-know concept that motions and interactions in natural media are of variation nature is confirmed. It is incorrect to study cycle structures without a corresponding medium of interactions. They should be studies in parallel. For this reason let us consider the spectrum of oceanic cycles. It is accepted to divide the oceanic regular cyclic changes into oscillations of certain periods. Thus, in (Plachotnik 1984) these changes are divided into ten groups: 1) micro-; 2) meso-; 3) daily; 4) synoptical; 5) monthly; 6) seasonal; 7) annual; 8) intra-century; 9) inter-century; and 10) glaciation periods.

Let us consider cyclic changes and phenomena of separate intervals, respectively.

1) **Micro-scale** cycles are determined by a kinetic viscosity of sea's water and represent turbu-lent pulsation of the smallest scale (0.3 mm to 1cm) and periods (0.1 to 100 sec).

2) **Meso-scale** cycles include surface gravitational waves, internal waves, lower boundaries of thermocline, and macrostructure of the ocean vertical stratification. Spatial-and-temporal scales range from several centimetres up to tens of meters and from several minutes up to tens of minutes. Initially it was incorrectly suggested that oceanic turbulence was spread all over the ocean, from its surface down to the bottom, but later it was found that it occurred only in the upper stratum, which was directory contacted the surface. In other strata of the ocean the turbulence does not play a significant role.

Table 8

Ocean Cycles' Spatiatemporal Characteristics

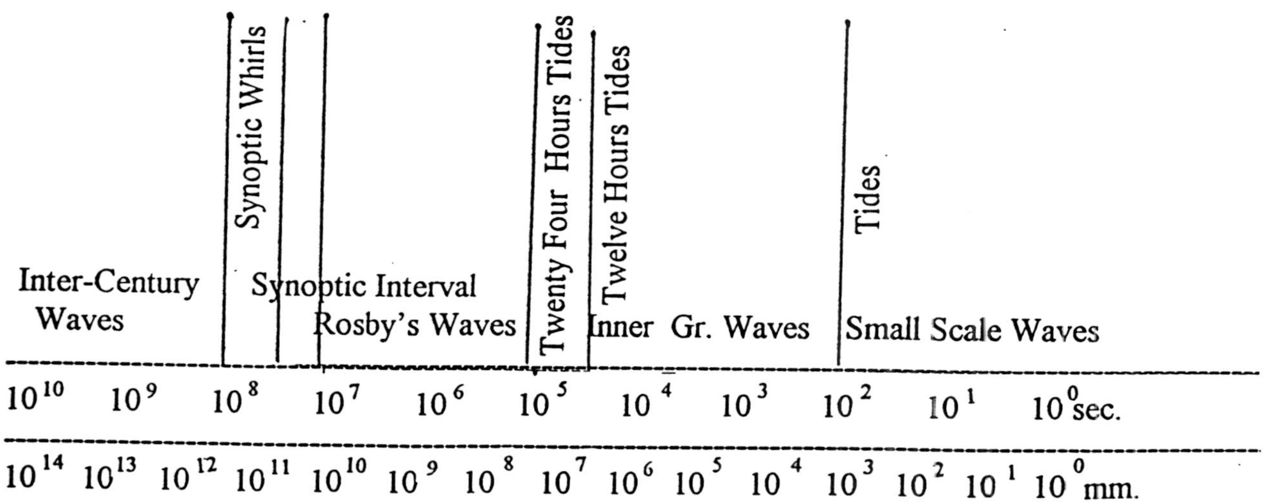

3) **Medium scale changes.** Spatial-and-temporal scales of these changes range from hundreds of meters up to tens of kilometers and from hours up to days. Oscillations of such periods include tidal waves and inertial motions of sea currents. The latters are mainly determined by the Earth rotation and are an important medium for energy contributions.

4) **Synoptic scale changes:** Spatial-and-temporal scales of these changes are higher by an order than medium-scale changes. They include free and synoptic vortices of the ocean, oscil-lations of sea surface level, and changes of the ocean upper stratum temperature.

5) **Large-scale changes.** Changes of this order include seasonal, annual, intra-century and inter-century changes. They are displacements of large-scale water masses, systems of oceanic currents and their effects in the atmosphere, and have spatial sizes reaching to sizes of hemi-spheres and the whole planet. Among them seasonal and half-year cyclic changes are studied best of all. Insufficient study of large-scale cycles is due to a certain degree to the fact that they are of all-planet sizes and nature and hence, it is difficult to distinguish spatial scales, which realise annual, intre-century and inter-century changes other than the Earth size.

Quantitative investigations of phenomena of the mentioned periods in the ocean are carried out with different depth and trends. Micro-scale changes are studied comprehensively and best of all. Then according to the depth of investigation meso-, daily and synoptic changes fellow. There are no considerations of the one entire cascade process of oceanic cycles. But what will such joint considerations give? First of all, the role and place of each cycle in division of common work will be clarified, possible limits of cycle deformations will be revealed and dependencies of different-order cycles on each other will be determined more accurately. The existing cascade considerations are rather of theoretical importance, or are carried out only within limits of definite intervals.

What are the specific features of the above mentioned spatial-and-temporal scales? First, the small scales are consequentially included into larger ones and displayed on their background. Second, for small hydrodynamic units the next larger unit is an external medium. Third, the spatial classification of structures corresponds to the similar temporal classification. The opinions of investigations of interaction problems are divided into two groups. According to the first one, micro-scale changes come into existence

3. Cycles' Significance

In the order to clarify the meaning of the ocean cycles, it necessary to reveal the role of its morphological structures in the oceanic division of works. The oceanic work's division is understood as a share of its contribution in establishing the globe stability. The ocean as a dynamic system, due to its characteristics such as mass, dimensions, density and other physical properties, interacts with a certain group in the general interval of excitations, which determines the share of the mentioned contribution. It means that the ocean is responsible only for neutralisation and dissipation of this group's excitation and it is realised through cyclic interactions of morphological structures and their reconstruction. Let us consider some specific features of the ocean's morphological structures.

It is convenient to accept surface, subsurface, intermediate, deep and bottom waters in the vertical distribution of the ocean water. According to considerations, directions of currents of intermediate and bottom layers are opposite, more precisely, north sub-tropical and sub-arctic anticyclonic turnovers of intermediate and bottom layers become cyclonic ones in the south layers. Instead of cyclonic turnovers of upper layers, the anticyclonic turnovers are dominating near bottom. And generally, everywhere in the ocean a current of definite direction in the upper stratum has its opposite corresponding current in the lower stratum. But what is the purpose and cause of such arrangements of motions is not clear yet. The questions of their origin are not also understood up to the end. According to Bourkov, surface and subsurface currents are of storm nature and current existing in the bottom layers are of thermocline nature, i.e. are formed because of difference in thermal and salt gradients. This event is based on the following facts. In the polar and sub-polar latitudes cold, heavy and fresh water are formed are formed because of domination of glaciation and precipitation over evaporation, these waters go down to depth of the ocean and are spread nearly horizontally occupying the ocean deeper layers, while equatorial warm, light but salty waters remain in the upper layer of their origin, which are moved to higher latitudes by storm current, and are replaced by deeper waters, and finally it results in great turnover.

Structural elements, which are in upper layer turnovers, for example, western near-coastal currents are also intrinsic to deeper turnovers. But the connection mechanisms of upper and lower turnovers are studied poorly, and the middle turnovers' physical nature are not clear at all. Similarly the data about deep and southern turnovers as well as vertical motions of the ocean waters are limited and not reliable. Only indirect considerations are available. Taking account of this fact, (Maksimov, 1970) truly notice that the basic problems of the ocean water nature and regime causes remain without a proper attention.

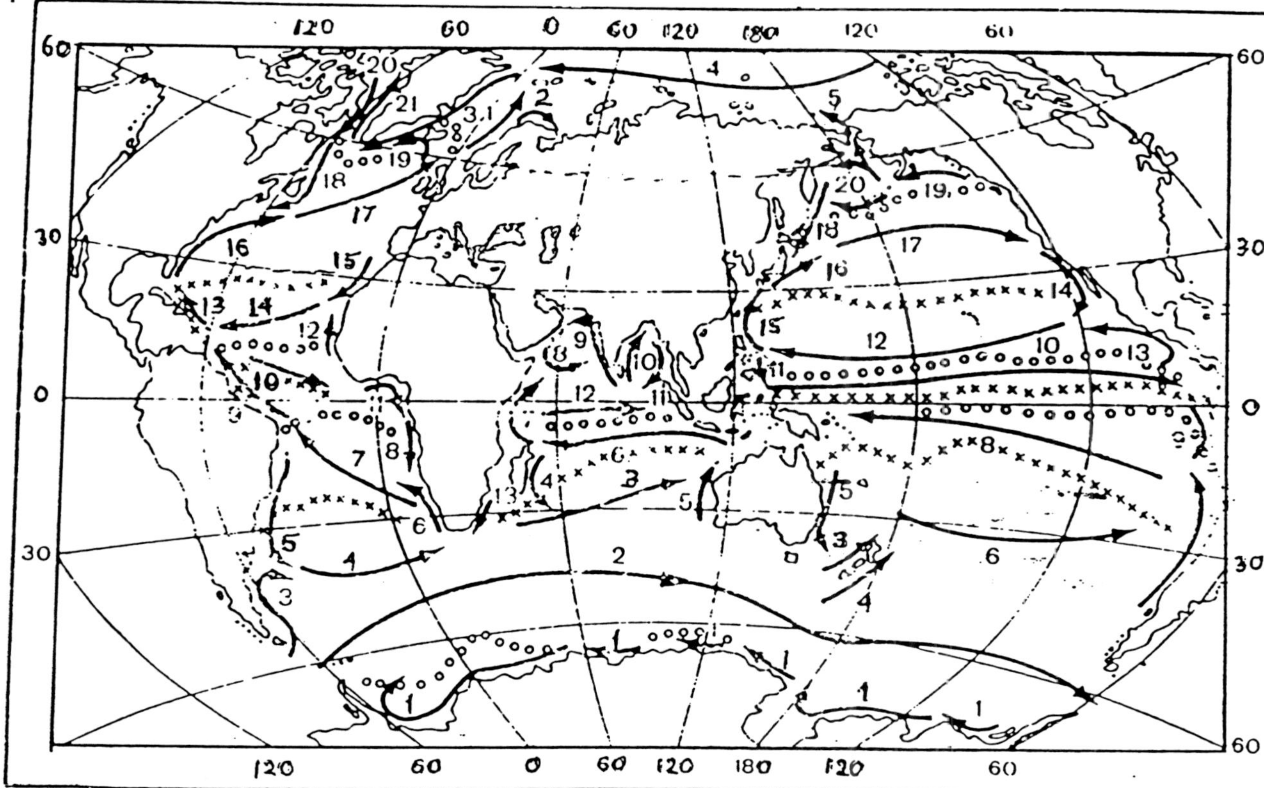

Fig.14 *Surface Large-scale currents and cyclic turnover of the World Ocean (after Bourkov)* ***Antarctica;*** *1. Coastal Antarctic, 2. Antarctic Circumpolar:* ***The Pasific Ocean;*** *3 Western New-Zeland, 4. Eastern New-Zeland, 5. Eastern Australian, 6. South Pacific, 7. Periodical Peruvian, 8. South Equatorial, 9. El-Nino, 10. Equatorial counter current, 11. Mindanno, 12. North Equatorial, 13. Mexican, 14. Californian, 15. Formosa,16 Kurpshio 17. North Pacific, 18. Gurl, 19. Alaskan, 20. Eastern Bering.* ***The Indian Ocean;*** *3.South Indian, 4. Madagascar,5 Western Australian,6. South Arabian,7. Somali, 8. Western-Arabian, 9. Eastern-Arabian,10. Western-Bengali, 11. Eastern Bengali, 12. Equatorial counter current:* ***The Atlantic Ocean;*** *3. Falkland, 4. South-Atlantic, 5. Brazilian, 6. Bengali, 7. South Equatorial, 8. Angolan, 9. Guienea, 10.Equatorial countercurrent, 11. Guinea, 12. Care Green,13. Atlantic, 14. North Equatorial, 15. Canary, 16. Gulf-Stream, 17. North Atlantic,18. Lablator, 19. Irningery, 20. Baffin, 21. Western Greenland;* ***Arctic*** *1. Norwegian, 2. North care, 3. Eastern-Greenland, 4. Western-Arctic, 5. Pacific. Circles show divergence, and crosses show convergence.*

Cyclic changes of oceanic water vertical turnovers are of special interest. Thus it is known, that in the equatorial area of the ocean there are the currents and are which exist under surface beginning from 50-100 meters down to 200-400 meters in a narrow belt with a width ±2□ and are directed opposite to the surface currents. This current in the Atlantic Ocean is called the Lomonosov current and has the following characteristics: the expenditure is **Q=30-40** svertroop, **V=150** cm/sec, the core oscillations are big. This current in the Pacific Ocean is called Gromvel's and has following characteristics:the expenditure is **Q=20-40** svertroop,**V=150** cm/sec, average oscillations are 500m.. The Antilles-Guanine counter current has the following charac-teristics: the expenditure is **Q=30** svertroop, **V=60** cm/sec, the width is 250 km, and the depth is 1000 meters. In the Indian Ocean this current has a seasonal nature and **V=60** cm/sec. The characteristics of direct and opposite currents are of the same order, but there is not a common opinion about their origin. Thus, according to some researchers, the counter currents, i.e. equa-torial and southern equatorial currents, are of storm nature. According to Stvertroop (Finlander, 1988), they are consequences of whirls. An answer to the question concerning origin of counter currents may be found by solving the problem of oceanic storm and thermocline turnovers interaction. Counter currents are also called balancing, compensation currents, i.e. such horizontal movements of water masses which compensate water deficit at some parts of the ocean and can be initiated in both surface and deeper, lower layers. Questions about their being primary or secondary, or supporting, currents are important but there being the joint cause of vertical turnover and interactions with horizontal ones are much more important. Sometimes separate branches of horizontal turnovers are also called compensating ones. Thus, in the Pacific Ocean the Kuroshio Current carries along equatorial waters to California region and according to Bourkov represents a counter balancing current of the south-eastern middle latitudes waters currents. In the opinion of some oceanographers, meridian currents that are subdivided into western near-coastal ones (the Gulf Stream, Brazilian, Kuroshio, etc.) and eastern border ones (Canarian, Bengali, Californian, Peruvian, etc.) motions connect zonal motions into one united system. But it is known that western near-coastal currents represent surface warm and salty ones and eastern near-coastal ones are currents of bottom cold and fresh waters. Hence, as it was already men-tioned, these turnovers of surface and bottom waters are united into one system. But why these turnovers should be initiated, which are the forces responsible for their origin, or what is the same, which forces are balanced, dissipated by these turnovers? It is unclear completely. Our knowledge about interactions of the oceanic turnovers of different scales and mechanism providing their steadiness, generally, is poor. And more over our concepts concerning nonsta-tionary reactions of the ocean in respect to change of external condition are insufficient. For example, up to now we do not understand mechanisms of interactions between wave motions and stationary currents (Monin A. 1974). Particularly, do transfers of kinetic energy shares have general direction or they are reversible?

Thus, there are "independent" turnovers in strata of the ocean, which interact with each other by means of whirls vertical motions and other mechanisms and average nonuniform distribution of the ocean's characteristics. In the other words they realise excitation dissipation. But turnovers are not the only mechanisms for smoothing and averaging the ocean's characteristics, there are many others. Actually the oceans have spatial-and-temporal characteristics of various scales. These characteristics are continuously affected by different-frequency excitations, which perturb and shift their initial values, giving rise to non-equal distributions. The ocean and its morphological structures, as all dynamic structures, are able to restore their equilibrium states. Without this ability to restore the existence of structures is completely impossible. Beyond doubts in the ocean the processes restoring its characteristics are all mixing, averaging and exchanging phenomena which are not permanent but are born and vanished every time during interactions with an external excitation (for a given scale.) Really all known geophysical phenomena in their essence are such restoring processes. They are: 1. The diffusion; (molecular); 2. The turbulence (of small scale);3. The diffusion (ordinary); 4. The convection; 5. The waves (stormy, internal, surface); 6. The whirls (free, of different signs, synoptic); 7. The advection; 8. The currents (permanent, temporary); 9. The turbulence (of large scale); and 10. The turnovers (of different signs and large scale). The mentioned phenomena do not give rise to non-uniformity and do not cause perturbations and shifts of the ocean balanced characteristics. On the contrary, they continuously balance, smooth and remove non-equal distributions of properties and characteristics and are the results of external excitation effects. If external influence is absent then the corresponding averaging and mixing phenomenon is also absent.

If a frequency of excitation is high, i.e. perturbations and shifts are local, then balancing and restoring processes are also local, that is they are of molecular diffusion, convection type. If a frequency of excitation is of medium scale, then restoring processes are of larger scale such as waves or whirls. And if perturbations and shifts caused by the external excitations are of large scale, then powerful processes balancing surface deformations of the ocean size order such as currents and turnovers are developed.

Thus the mentioned geophysical phenomena everywhere within the cycle balance unequal distributions meteorological and hydrodynamic characteristics of different strata and currents of the ocean and establish equilibrium within the limits of corresponding scale. It should be noted that the above mentioned phenomena are corresponding to certain Lorenz space- and-time scales beyond which they are not the efficient smoothing mechanisms already. Thus, if an averaging is made by means of diffusion, then the ideal condition for averaging and balancing is $\mathbf{L} \ll \mathbf{D}\tau$, where **L** is a linear dimension of the system, **D** is a diffusion coefficient, and τ is a characteristic time of the process. Beyond these limits the role of diffusion in averaging processes is not substantial. Such conditions for characteristic time and scale also exist within the frames of other phenomena. Such behaviour of processes at some scales may be caused by a value of the excitation exceeding a critical value and result in a chaos. Concluding the said above, it can be noted that geophysical cyclic phenomena and morphological structures are not of accidental nature. Their interactions are quite determined and display organisation and particular directions of transfers and exchanges. They serve as balancing and dissipating elements for external excitations (for a given scale) and never been the same on the geological time axis.

4. Oceanic Currents and Fronts (Linear Structures)

Among the morphological structures of the ocean there are currents (fronts, layers) which are related to a significant share of cyclic changes of the ocean. Usually a front or a border is understood as limits of water masses of different scales, where their main physical and chemical properties and characteristics (temperature, salinity, pressure, speed, rate of mixing, etc.) are subjected to drastic jump-like changes. They are realised within spectrum of main elements of turnovers, motions, beginning from one meter up to the dimensions of the whole ocean, from the surface down to the bottom. The oceanic fronts are observable even from the cosmic altitudes, and they are outputs and traces of deeper fronts on the surface. There are many invisible to an eye border within the depths of the ocean, many of which never contact the ocean surface. Fronts separate, warm and cold waters, fresh and salty waters. If cold or salty waters are situated on the one side of the front, then the other side, there are warm or fresh waters, which may result in sufficiently large and stable gradient along the width of the front. Hence a steady state of front is due differences of currents (and other properties) between the two sides of front, and substantial decrease of a speed.

Thus the ocean water is everywhere separated by numerous natural boundaries, into different visible and invisible sites, layers and layer groups. Some of them may be called water masses, and this name is hardly suitable for others because these layers and their groups are very small. These groups of layers, fronts being an integral one by their structures are quite different in their importance. Space-and-time subordinated structure of external excitations determining the similar structure of the World Ocean is corresponding to such "requirements". And forced "division of works" gives rise to its horizontal and vertical stratification of different scales, non-uniform distributions of its physical-and-chemical characteristics. The existence of subordinated scales of fronts should be taken into account mandatory when considering the ocean structure (Fedorov, 1983) .

As it was already mentioned, the important feature of fronts is preservation of spatial gradients in strained state for a long time, on the contrary to permanently existing averaging and mixing processes (turbulence, diffusion, convection, and so on), which are aimed to their demolition. This self-preservation property of fronts (boundaries) is due to both reverse flow of energy and specific features of local water thermodynamic characteristics and zones of relatively high gradients (Fedorov,1983). Really, a front is stable not for the fact that turbulent mixing, diffusion or other processes balance different values and properties of water masses, but for the fact that there always are permanent reverse flows of energy which feed and preserve this difference of gradient. Just in the same way, different-scale structures are stable not for the fact that convective, turbulent, whirl and other movements do not balance, equalise corresponding gradients, but there is a permanent inflow of energy within the limits of a given scale, which supports jumps and drops of physical-and-meteorological values. Thus, for example, the fact that the waters of the Mediterranean Sea are able to cross the whole Atlantic Ocean and do not loose their physical characteristics evidences that fronts between them and ambient waters remains always in sharp, strained state. The same may be said about the salty waters of the Red Sea which penetrate the Indian Ocean and do not mix up, go for long distances, preserving their physical values and charasteristics. All these are consequences of long-action and constantness of forces giving rise to them. The main characteristics of fronts are given below.

Typical scale characteristics of the fronts in the World Ocean
(According to the observation data) (after Fedorov, 1983)

Maximum width of gradient zone	10m to 10km
Temperature drop in this zone	1^0C to 6^0C
Salinity drop in this zone	0.2 to $10^3/_{00}$
Density drop in this zone	10^{-1} to 10kg/m^3
Temperature horizontal gradient	0.1 to 30.0° C/dm^3
Salinity horizontal gradient	0.1 to $10^3/_{00}$
Density (per 1km) horizontal gradient	10^{-1} to 10kg/m^3
Inclination of front	10^{-1} to 10m/t per 1km

According to Fedorov 1983, the following conditions and sharpening of gradient strains within fronts: rotation of the Earth and forming of internal layer, involving turbulence, inertia (non-linear) effects, forming of compression in the course of motions, and vertical motions. It is known that strong changes take place within fronts and they may "travel" on long distances, to get stronger, or weaker, to be spread out and restored again. In the process they may be born and exist without direct effect of wind. Oceanic fronts, wave oscillations generated within them, whirl excitations and movements, fine thermal structures are playing an important role in cascade scales of kinetic energy, enstrophy (nutrients): in transfers from one part of the spectrum to an other, to dissipate scales, in processes of conversion of potential energy into kinetic one and vice versa. But one important feature of front is not paid enough attention: it prevents oceanic waters to be spread out in random direction by uniform currents, in other wards, it prevents random propagation of excitations. More high is stratification of medium more hindered is propagation of excitation in it. Hence, the stratification in the ocean, as partition in organisms, plays the same role, a protection. Preventing spreading of excitation and dissipating it in upper layers, it protects stability of a system.

The effect of the Earth rotation on the fronts of the ocean and the atmosphere is uniform. Thus, movements of oceanic fronts necessarily give rise to shifts of atmospheric centres arrane-gment, changes of stream current located over the ocean. Then changes, in fronts of the northern area of the Atlantic Ocean, result in complicated and significant movements of the Iceland atmospheric pressure minimum center. The Gulf Stream, as rhythmical change of frontal separations and activity, give rise to the similar change of the Iceland center position. 19-year rhythmic movement of inclinations is also observed in the changes of position of the Iceland atmospheric pressure minimum (Maksimov, 1970).

If front scales are sufficiently large and well distinguished in density field, then a geostrophic balance of motions and masses determined by velocity and density values are observed on the both sides of the front. In these case fronts has some inclination. The tangent of inclination is proportional to drop of velocity along the front:

$$tg\,\alpha = \frac{f^* \rho(u_1 - u_2)}{g(\rho_1 - \rho_2)}, \qquad (1)$$

where **g** is the acceleration of gravity, **f** is a Coriolis force, $\mathbf{u}_1$ and ρ_1 are the current velocity and light water density along the front length, $\mathbf{u}_2$ and ρ_2 are the velocity of current and heavy water density on the other side of front. This formula called the Markules formula results in regular observation of double whirl existence in oceanic, as well as atmospheric fronts (Fedorov, 1983). According to Rosby, the basic similarity of the ocean and the atmosphere is the existence of vertical motions giving rise to fronts in them, particularly lowering of cold water and air masses in stream current zones. The largest separations or boundary lines of the World are its well-known currents.

Famous Currents of the World Ocean [Ocean, BSE]

Atlantic Ocean

Warm currents	Cold currents
1. North Equatorial Current	1. Bengali Current
2. South Equatorial Current	2. Canary Current
3. Equatorial Counter current	3. Labrador Current
4. North Atlantic Current	

5. Gulf Stream, **v=2.5m/sec, h=1000m**
L=100km, Q=75 svertroop

Indian Ocean

1. South Equatorial Current
2. Monsoon Current

Pacific Ocean

1. North Equatorial Current
2. South Equatorial Current
3. Equatorial Counter current
4. Kuroshio **v=1.7m/sec, h=500m**
Q=55svertroop L=2000km

South Circumpolar Drift

Q=200 svertroop (1 svertroop=10^6 m^3/sec).

While the today's developments of the questions of origin, formation of warm currents and cyclic changes of their characteristics may be considered as sufficient, then the questions of origin of cold currents are completely unclear.

Thus the ocean is not a system lacking internal structure, i.e. a homogeneous system. It is multy-layered everywhere and nearly in the whole depth by boundary lines and fronts and endowed a certain self-organisation. Distribution of layers is subordinated and stable and different layers have different roles and contributions in interactions of different scales.

5. Oceanic Cyclones and Anti-Cyclones (Non-linear Structures)

Whirl movements of the World Ocean, cyclones and anticyclones, are less widely spread that, the similar motions of the atmosphere and no less importance for its structure and interactions. For the first time existence of such motions in the ocean was confirmed 40 years ago, and their nature is well studied now. But nevertheless, distribution of whirls, questions and problems of their origin, significance and interaction scales are far from being satisfactory.

It is know that lifting of oceanic deep cold and fresh waters to upper layers is observed in central parts of cyclones, and opposite process is observed in anticyclones,i.e. surface warm and salty waters are lowering down to bottom layers of the ocean. Such motions occur not only in upper layers of the ocean, but also in bottom layers and, according to some sources, are connected to each other. It is found that the velocity of current in whirl field is different at different depths, e.g. at 200 to 300 m it is up to 25 m/sec, at 400 to 600 m it is 35 cm/sec, at 1000m it is 20cm/sec, and at 1500m it is 10 cm/sec. But in the course of motion of whirls deep currents preserve directions of their motions. According to the data of Stida, an average velocity of oceanic surface waters is 10 cm/sec, and that of whirls is higher by one order.

There are several viewpoints for whirl's origin, but neither of them become a theory yet (Fedorov, 1983). In 1970s an assumption of "dense location" of whirls in the ocean was set forth, according to which the whole ocean is composed of and filled with whirls, what means that there is the continuous force giving rise to them. The causes of whirl origination are kinetic energy, baroclinic instability of large currents, gradients of various natures such as termoclinity, baroclinity, etc. (Bourkov, 1980).

As atmospheric whirls the ocean whirls also have their life cycle: origination, development, the stationary state and decay stages. They interact with permanent currents, fronts, internal waves, bottom relief and each other, even born the likes. At the same time there are the differences between them. Thus, axes of atmospheric whirls are mostly fixed and only their frontal areas are moving, hence motion is of wave nature. Except for wave nature, which are called synoptic or free whirls, oceanic whirls may be of advective nature when both axis and internal content of whirls are moving. These later whirls are called frontal whirls. As the observations of "Polygon-70"showed. displacements of free whirls were dominantly of wave nature. They, on the contrary to frontal single whirls, are distributed in chess succession, that is cyclone-anticyclone alternation. Whirls of this type have much longer life and are moving much slower than atmospheric synoptic whirls. They are occurred as highly energetic excitations which space-and-time scales are changing within a very narrow band of the spectrum (Monin, 1974).

Frontal partitions and boundary lines (Gulf Stream, Kuroshio, etc.) are zones of origin of different-scale and different-sign rings where whirls of different signs are born, and which as a whole are slowly, with a speed of 3 or 4 km per day, moving in the ocean. Thus, in the field of the Gulf Stream motions approximately five pairs of cyclone-anticyclone are born every year, cyclones being located on right of the current, and anticyclones on the left. Well-developed cyclones have a diameter approximately 2000 km and penetrate down to the ocean bottom. Their rotation speed, in upper part reaches up to 2 m/sec and sometimes more. Whirls motions are mostly of transient nature, they are rather of drift than wave nature. Some cyclones may exist up to two years, transferring gradually their rotation energy to environment and may travel from the place of their formation a distance over 1000 km. Just in the same way strong whirl motions are originated in the Kuroshio current. Anticyclones are formed on the right of it, and lower surface warm and salty waters down to lower layers, and cyclones are formed on the left side, and lift cold and fresh waters from the bottom up to surface layers. Classification and characteristics of whirl motions are given in Table 9.

The table shows that whirl may have various natures and various causes of origin. Some whirls have isolated cores (lenses) which may form just from the beginning, or may be originated by means of whirls, or may be captured during whirl travelling. Processes of formation and decay of the mentioned areas, particularly, what is the mechanism of gradual accumulation or, on the contrary, transfer of energy to ambient oceanic water, are of great interest for the theory. It was found that cores contributed greatly to motions of thermomass along the fronts of origin of local fronts (Fedorov, 1983).

What is the role of whirls in processes of cascade transfer of kinetic energy and hydrophysic fields changes? According to Malgus, whirls, independent on their nature, scale and rotation direction, transport portions of accumulated energy to other place and make balance quicker than other motions of energy (diffusion currents, convection, etc.). Whirls, and especially turbulent whirls, transfer energy at the expense of kinetic energy of rotating mass, independent of the fact that they are caused by heat convection, instability of current, or some other mechanism. Free whirls play a significant role in processes of energy transfer and conversion due to their high energy content, particularly, in changes of oceanic upper layers' water temperature, which each up to several degrees, for distances of tens and hundreds kilometers. Energy of whirl current exceeds the average energy of oceanic current by a factor of 2 for strong currents and by a factor of 20 to 40 for weak ones. They contribute greatly into thermomass exchanges occurring in a meridian direction (Menabe and others, 1972). Taking account of these specific features of whirls, it should be suggested that they play an important role in establishing equilibrium within the limits of their scale.

Table 9

Whirls of the World Ocean (after Fedorov)

No	Type	Example	Rotation direction	Longitudinal section ,km	Layer thickness,	Life time months	Motion direction	Motion km/day	Matching potential energy , joul
1	2	3	4	5	6	7	8	9	10
1.	Ring	Gulf Stream Kiroshio	A. and cycle	100 - 300	0 - 4000	2 - 24	south-west	2 - 8	10
		Atlantic Circumpolar	A. and cycle	150 - 450	0 - 400	2 - 24	north-east	0.2 - 8	130*10^5
			A. and C	100 - 180	0 - 3000	?	north-west nort-east north-east	2 - 3	?
2.	Boundary whirls	Gulf Stream Shelf fronts	C. C.	20 - 60 20 - 40	0 - 100 0 - 100	0 .3 0.1	along the current	30 - 60 1-2	? ?
3.	Core-less synoptic whirls of open ocean	Whirls in the region of "Polygon-70"and "P-70"	A. and C.	100 - 200	50 - 2000	1 - 12	dominantly to the west	1 - 5	(1÷2)*10^5
4.	Cored soptic whirls of open ocean	large	A. and C.	80 - 150	0 - 2000	> 24	north	5 - 15	10^{15}
		small	A. and C.	10 - 30	100 - 2000	> 24	north-west, south	~10	10^{11} - 10^{12}
		arctic	A. and C.	5 - 30	30 - 350	?	?	?	
5.	Whirls which transform oeanic wather and whirls	eastern Australian	A.	200 - 250	0 - 2000	2 - 3	west, south	< 1	?
6.	Convective whirls	Region of the Weddel se	C.	10 - 30	0 - bottom	?	?		?
7.	Whirls originated from s	Wesern Pacific Ocean	C.	100 - 200	0 - 1000	> 2		?	(1÷10)*10^5

Is there any relation between whirl motions of the ocean and the atmosphere? According to Lebedev, centers of oceanic and atmospheric whirl motions are not coinciding. On the country to that it is noted in Maksimov that centers of oceanic and atmospheric whirls are coinciding. The solution of this question is very important from the viewpoint of the theory, because in the first case, it means that whirl motions in the ocean and in the atmosphere are independent on each other, and in the second case, it means that they interact with each other and in some events they have planetary causes of origin, which is not lack a basis, when taking account of the existence of critical triangles in the Earth's crust. Now enough data and facts are gathered allowing to say that there are also whirl motions and displacements of different scales in the Earth's crust. According to geological investigations, they become the causes of perturbations in the poles' positions, changes and inversion of the Earth magnetic field, continents' drift, magnetic contributions, regularities of orogenesis, and changes in the Earth's shape and its rotation angular speed. But today's level of investigations of spiral formations in the Earth's crust is far from being satisfactory and does not correspond to the scale of their spread.

6. Generation and Formation Causes

For correct understanding of causes giving rise to oceanic cycles of different scales and regular changes a keystone is investigation of forces and factors giving rise to its large-scale changes. Medium- and small-scale changes, according to ideas existing in hydrodynamics, come forth from large scale instabilities through cascade transformations. But their formation is affected by independent causes too. Large-scale changes of the ocean include its large-scale turnovers. Let us consider the viewpoints concerning their origin.

There is no united opinion among specialists about the mechanism and causes of origin of large-scale water turnovers in the World Ocean.

In the opinion of some specialists (Svertroop, Stomel, Mank) these, turnovers are caused by direct effects of tangential friction of storm whirls. Thermal and salt exchange processes (thermocline, barocline) have not a significant role. Strain of currents observed in the western coastal regions is determined by the effect of $\beta = \partial f / \partial y$. Thus, according to this point of view, surface currents are completely determined by storms, and gradient currents are mainly of thermocline nature.

In the opinion of the second's group researchers, the basic causes of the world ocean turnover's origin are processes of thermal and salt exchanges. According to this point of view storm currents do not represent turnovers of ocean upper layers but support their formation. In the opinion of these researchers the strain of currents observed in the western coastal regions is not determined only by the effect of β, and a joint influence of the ocean bottom relief and baroclinity play a role of the same β-effect. In other words, nonuniform thermal and salt distribution and their unbalances constantly cause large-scale turnovers. In the regions of subtropical anti-cyclones heat inflow is due mainly to a positive balance of heat, and in the regions of sub-polar cyclones by a negative balance.

Another two mechanism of turnover's origin may be stated as follows: 1. a mechanical version of motion origin: in free ocean influence of tangential friction of wind and β-effect cause, at first, meridian currents, which are directed to the north and to the south, strong meridian compensation (what?) currents formed at medium latitudes and in the western coastal regions being consequences of the former ones; and 2. thermal version (Sarkisian et al.): in free ocean nonuniform heating of its surface causes, at first, origin of left currents which are directed from the east to the west and which give rise to strong, straighten currents of the western coastal regions due to presence of meridian boundaries, density advection and β-effect. Thus, in the case of the first version, the principal forces giving rise to turnovers are the meridian current causes by storm effects, which then result in of belt currents. And in the case of second version, belt currents, considered important and primary, are caused due to nonuniform heating and then they give rise to meridian currents. In the result these currents of both directions are united and form turnovers. In the opinion of another group of researchers both mentioned factors are simultaneously and are important in forming of turnovers.

Another group of researchers tries to substantiate interactions between oceanic water turnovers and structures by means of geostrophic models. Th geostophic models (according to their definition) involve only by the ocean's upper layers, though. The deeper turnovers, that also contribute to total oceanic motions, are not described by these models and therefore they are not complete. Geostrophic models allow only make a conclusion that the total turnover of layers of the World Ocean is predominately of storm nature.

According to Bourkov, the total horizontal turnover of the World Ocean is composed of two systems, upper and lower ones. The physical nature of the upper turnover is thermal and stormo-genic. In this sphere the ratio of storm and thermal energies' expenditures is 2:1, and the mean temperature of water is higher by 3.8°C than the mean temperature of the ocean. In the lower sphere the physical nature of turnover is only thermal, its energy expenditures are lower by a factor of 3 than that of upper layer, and the mean temperature of water is higher by 3.8°C than that of the ocean. The interface between them is located at the depth of approximately 1500m and they are connected via vertical time. In the opinion of Burkov among these turnovers the leading one is the lower sphere (proceeding from the estimates by Reid) and the upper sphere is adapted to the lower sphere turnovers through the mechanism of geostrophic divergence. But what are the forces responsible for the lower sphere turnovers it is not clear yet. More over, in the southern hemisphere these spheres are united and form huge Antarctic circumpolar current with the following data: 19000 km in length, 3000 km in width and 3 km in depth, i.e. down to the bottom of the ocean. This current being the greatest among the oceanic currents has no clear theory about its origin nature. According to some sources it is formed by thermal and storm factors.

According to other authors increase of surface water temperature and salinity (due to evaporation) or their decreases (due to precipitation) causes lowering (or lifting) of these masses (due to the Archimeds force). This forms a basic cause of thermal turnovers.

Oceanic turnovers, beyond the doubt, are related to its surface deformation and nonuniform distribution of other characteristics in the ocean. And, hence, it is obviously incorrect to connect the question of turnover origin only to thermal excitation by the Sun, as well it is incorrect to exclude it completely. It should be assumed that besides the thermal influence of the Sun on oceanic turnovers, more powerful sources of excitation affect them: the whole system of tides, nutation, rotation, and other multiyear period exogenic forces.

Taking account of this fact some researchers use these sources of energy when substantiating questions concerning ocean turnover's origin. Thus, according to Maksimov, the oceanic turnovers are directly connected to its surface level deformations, which occur always under the influence of tides and periodic forces. Effects of tides and all exogenic excitations on the ocean are complex in their nature and have different periods. A general influence of the sum of these forces on the ocean is sufficiently complex and of irregular nature. In the result the level of the World Ocean surface permanently oscillates according to a complex law. At the first place these oscillations are displayed in large currents of the ocean and create a specific system of rhythmical changes, including not only alternations of potential and kinetic energies transfers of large-scale water masses but also rhythmical changes of all other characteristics (temperature, density, pressure, salinity).

Questions of tides and their transformations are among the most complicated, up-to-the-end-unsolved questions on the Earth. Despite this the potential force of tide was expanded into a series with harmonic terms in 1921, and a total number of the series terms was found: 389, their particular period being: 99 long periods, 157 daily, and 129 half daily. Nevertheless many questions concerning interactions and dissipation of the mentioned terms are not clarified and understood up to now.

Long-period terms of tide force, being typical representatives of gravitation force, are proportional to a longitudinal terms $(1-3\sin^3\varphi)$ and, hence, give rise to complicated non-periodically changing components which contract the ocean shell and cause changes in a position of the ocean surface level.

The deformation of the World Ocean surface level positions is determined by the following sum:

$$\Delta Hz=\Delta H1+\Delta H2+\Delta H3+\Delta H4+\Delta H5+\Delta H6+\Delta H7+\Delta H8+\Delta H9 \quad (2)$$

where **ΔH1** is the Moon half-day tide;
ΔH2 is the Moon daily tide;
ΔH3 is the Moon's 19-year declination tides;
ΔH4 and **ΔH5** are the deformation forces caused by free and forced motions of the Earth's poles;
ΔH6 is the deformation forces caused by changes in the Earth rotation;
ΔH7 is the Sun half-day tide;
ΔH8 is the Sun yearly tide and
ΔH9 is the influence of gravitational force.

Among these sums the determination of **ΔH9** is very important. In order to estimate it the following expression is assumed:

$$\Delta H9= -1/6(\omega r/6)^2 (1+K-h)(1-3\sin^3\varphi)\Delta g \quad (3)$$

From this formula follows that abnormalities of gravitational force are distributed nonuniformly on the Earth. Thus, the forming of the surface level deformation of the World Ocean, is subjected to two main forces: the first one is the gravity, which is nonuniform in different regions of the Earth, and the second one is tidal force caused by motions and activities of the Sun, the Moon and other heavenly bodies. The balance of these two groups of forces, i.e. the position of ocean surface level, is not difficult to find from mean space-and-time values of the mentioned forces. However, such an approach mask, question of forming and interaction of different scale cyclic changes, though individual and separate considerations of which are sometimes much more important.

Let us consider the large-period terms of tide and the phenomena caused by them. Their contributions into the total tide are different and have the following form (Table 10).

Table 10.

Basic Component of Long-Term Changes of Tidal Force (after Maksimov)

Period	7.1	9.1	13.6	14.8	27.4	36.8	121.8	171.8	182.6		186	1805	160
Contrib%	11.63		41.2		22.31			14.24			10.71		

Following changes with the mentioned periods are observed in the ocean.

The first group represents oscillations of sea level with periods from 7 to 8 days, changes of their currents, atmospheric pressures and turnovers.

The second group represents 14-day oscillations of surface and its currents. This is the period of changes of the ocean internal waves, motions of the Earth, poles, magnetic field and seismic activity.

The third group. 30-day changes of sea surface and its currents, and monthly changes of internal waves, characteristics of the Earth magnetic field, and frequency of earthquakes are observed.

The fourth group. There exist: **a)** half-yearly changes of the ocean internal waves; **b)** tides of the same periods in the ocean and the atmosphere; changes of the total turnovers of the atmosphere; **c)** oscillations of sea surface and its turnovers lasting nearly one year.

The fifth group: a) changes of water temperature, as well as changes of the atmospheric pressure and the mentioned periods of the atmosphere turnovers are observed; **b)** the Soros cycle: coincidence of alignment of the centres of the Sun, the Moon and the Earth in one line, and of monthly and half monthly tides stages; **c)** the Peterson's cycle: syzygies recurrence of aphelion and perigee, highest cycles of tides, etc. The mentioned long-term terms of tide strength give rise to different deviations with respect to the mean level of ocean surface, i.e. deformations (**ΔHz**). Therefore, there is a question: what are the mechanisms restoring the mean ocean surface level during tide cycle, or, in other words, how values are equalised, smoothed; what are the mechanisms of ocean level balancing? And finally, how tide strength is transferred from the Earth's crust to the ocean and the atmosphere, and vice versa? These questions are not superficial ones best require deep investigations. According to Perstev and Ivanov, large semiaxis of horizontal displacement ellipses, for the Earth's surface points, may reach as large values as 4mm even in the centre of the Asian continent under the influence of the ocean tide ($\mathbf{M_2}$ wave), this value amounts 10 percent of displacement in the meridian direction caused by the Earth tides. According to Maksimov turnover to oceanic waters are of deformation nature, even if not the whole but part of them, i.e. they are born due to deformations and are directed to eliminate and balance these deviations.

Thus, large-scale turnovers of the ocean, according to some researchers, are arisen in respond to its excitation as a whole and are dissipating elements for tidal energy, regulating through these events a stability of hydrosphere as a whole. It is possible that tidal energy may be coversed into internal waves and turnovers, be transferred to the Earth's crust and/or the atmosphere, etc. This point of view about origin of large-scale turnovers, despite being a perspective one, has unclear aspects. At first, a particular component of tidal strength long period is not clear, and since it affects, as gravity force, all layers of the ocean in the same manner, then it should be expected, that turnovers are to be propagated down to the ocean bottom and involve its mass as a whole, though this effect is not observed. On the contrary, turnovers of different signs (directions) are observed at different depth, question of their being united or counterbalancing is not developed yet. It is not clear, should these period be attributed to specific features of coastal zones, basins or bottom relief of the ocean, or to seismic excitations of the ocean bottom, existence of high gradients in rift zones, or in order to find their origin it is necessary to involve another group of conditions and forces (say, rotational, nutational, etc). Anyway, the above-mentioned causes, apparently, are not all, which are responsible for the ocean surface deformation origin. Other exogenic interactions such as the Sun activity, its bowls dynamics and other are also contributing here. Thus, according to (Maksimov, 1970), the ocean receives sometimes powerful impacts and shocks which are distinctly expressed in surface deformation of the ocean level. Just in the same way, 11-year oscillations of its mean level are observed everywhere the absolute amplitude of this oscillations does not exceed 4 to 5 cm. These oscillations are one of the

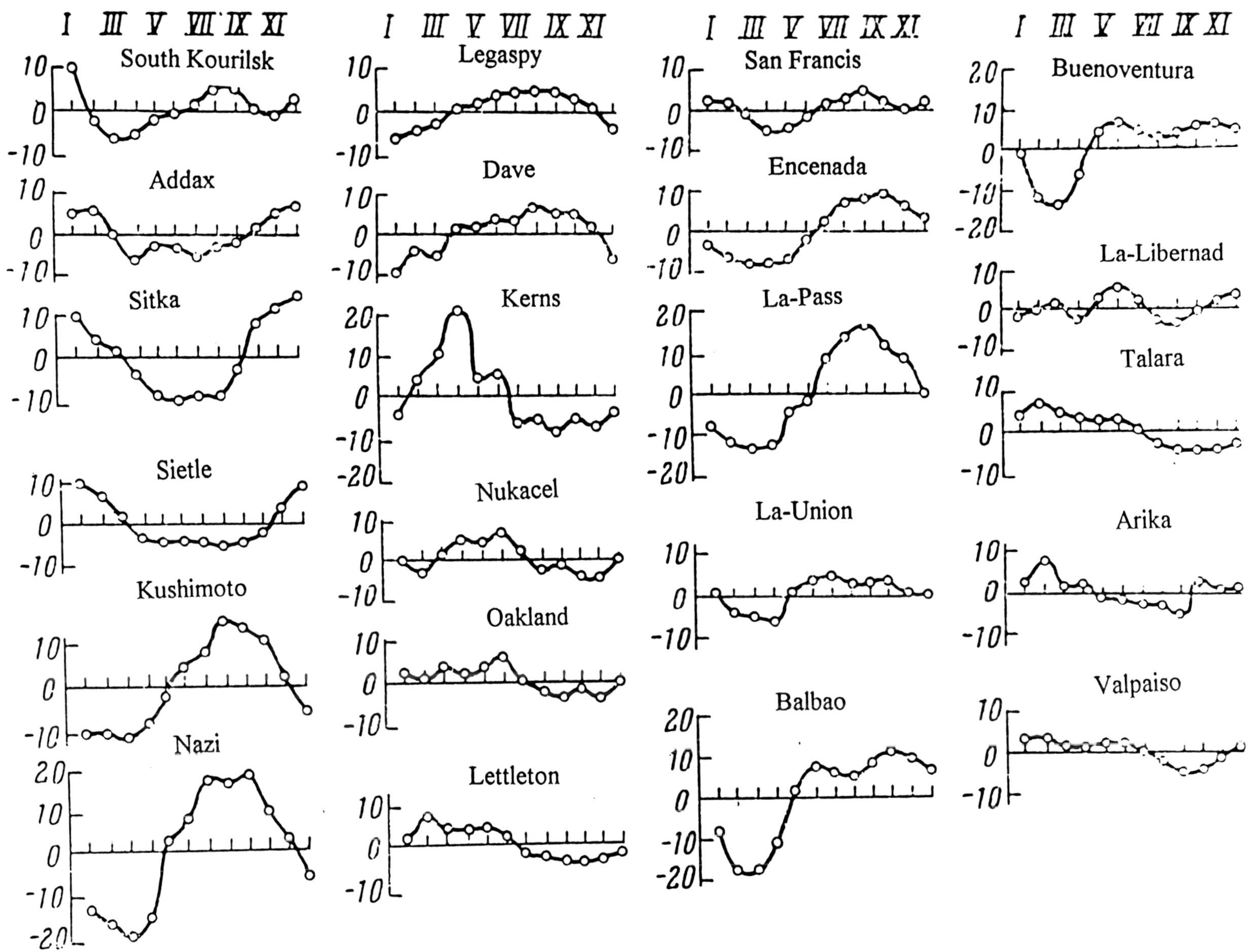

Fig. 15 *World Ocean level in the different points of its surface (after Maksimov)*

main causes of origin of long-term changes of the ocean level. According to Maksimov, 11-year cycles in the behaviour of the ocean level and the changes of the same periods of current which are related to the former, are caused by direct influence of 11-year cycles of the Sun, and, as some researchers are noted, are not consequences of 11-year rhythms of the atmosphere.

Contributions from rotational (may be also nutational) forces to the problems of formation of the ocean's large-scale turnovers, which may be promoting and inhibiting, should not be neglec-ted. But their estimations are much more complicated and fine problem, which requires many details of daily cycle of true ocean angular moment, particularly, its beginning and end, magni-tude, change, maximum, structure, relaxation and dissipation. Magnitude of relaxation with respect to rotation and nutations is especially important, during which the ocean transfers a portion of its angular moment to environment: the Earth's crust, the atmosphere, transforming and forming its morphological structures till the next portion should come. Or it seems that large-scale turnovers are existing forever, independent of these excitations, have no start and end, are not changing from cycle to cycle, do not take part in dissipation of rotational energy. But really the pattern is quite different. The effects of these acting forces on large turnovers are existing, waters at different depths from the ocean level are constantly balanced and smoothed, and a balance within cycle limits is established due to them. Then, an and its morphological structures as dynamic systems posses definite energy and relaxation ability, which are reasons why they do not respond to external excitation instantly. Nonresonance reactions of the ocean, including also stationary effects, are related to internal domains of frequencies, which require a particular time. A particular time is also required for processes eliminating internal instabilities. Turnovers have their own periods and, hence, resonant properties and capability to generate oscillations. Observed continuity of motions masks cyclic recurrence of affecting forces, and is consequence of the fact that medium is "working" with its so called maximum rhythm. When an influence of one cycle is propagated in it directly, and the next is following without interruption, when the last stage of relaxation of one cycle is coinciding with the starting part of excitation state of the next cycle. Taking account of it, let us discuss in details one of the probable forces forming the ocean large-scale changes. Let us briefly consider existing concepts of the Earth rotation angular moment. It is known that an absolute angular moment of a unit mass with respect to the Earth rotation is determined by the following relationship:

$$\mathbf{M=\Omega a^2cos^2\varphi+Uacos\varphi} \qquad (4)$$

where $\mathbf{\Omega}$ is the angular velocity of the Earth rotation, $\mathbf{a}$ is a mean radius, φ is a latitude, and $\mathbf{U}$ is a latitudinal component of the velocity (is positive in western direction). Often the first term is called ê moment, and the second term - a moment of relative angle.

A balance of angular moment for a unit volume is determined by the following equation:

$$\rho \frac{dM}{dt} = - \frac{\partial p}{\partial \alpha} + F_{\lambda} \cos \rho , \qquad (5)$$

where ρ is density, **t** is time, λ is longitude, $\mathbf{F}_\lambda$ is latitudinal component of friction force (is positive in western direction). This equation shows that the changing rate of absolute angle moment contained in a unit volume is equal to a sum of pressure and friction forces affecting this volume.

Obviously, the rigid Earth globe plays a definitive role in counterbalance of the angular moment (4). The contribution of the ocean and the atmosphere to Eq.(4) are considered as insignificant taking account of their masses. But not so insignificant that their inverse effects on the rigid globe angular moment could be neglected. These effects were observed for the first time in 1950s, when effects of the total angular moment of the atmosphere were recorded in seasonal changes of the Earth globe angular moment. It was found (Menabe and Brain, 1972), that corre-lative strong negative relation existed between them and transfer of angular moment was meridian and was carried out through the ocean. There also exist a constant transfer of the ocean angular moment to the rigid globe, which, in this case, is latitudinal and carried out through the difference of ocean level at western coasts (Oort, 1988). But how these transferred moments are further distributed and transformed in the Earth's crust and which, as it was already mentioned, even in the centre of the Asian continent are equal to 10 percent of own tidal displacement of the Earth's crust, is completely unclear. According to some researchers, it is only known that continental forces connect large-scale motions of the ocean and atmosphere to particular displacements of the Earth's crust. Apparently a solution of this question would help to clarify the Earth's crust strain field (arisen due to angular moment), to predict its structure and manifestations of earthquakes in zones of inclination. This concept could not be propagated for large scales of time because strains of the ocean and the atmosphere (arisen due to angular moment) and their changes are lower by, an order than those forces which regulate daily processes of the Earth's crust. Rotations of angular moment among the Earth's crust, the ocean and the atmosphere, its size, directions of transfer and transformations generally are quite unstudied today. While currents and turnovers of the ocean and the atmosphere transfer the angular moment with approximately daily period to medium latitudes above the equator and in latitudinal direction, then in the Earth's crust it has an opposite direction: from medium latitudes to the equator and is connected to motion of masses along lengths of inclinations. The inclinations have north western and south-eastern directions in the north hemisphere and north-eastern and south-western directions in the south hemisphere and occupy definite latitudes (Oort, 1988). It is clear that this opposite behaviour of angular moment in plates a h s not daily period and is accumulated, which reaching a definite level is necessarily transformed into earthquakes in the mentioned latitudes of ±40 to ±60. Such a transformation possibility of angular moment being considered as natural one, nevertheless is not confirmed yet by data and observation statistics. Even most general qualitative ideas concerning daily accumula-tion of angular moment and regularities of its further transformations are absent. Without doubts, there are completely regular geophysical and independent on each other process (cycles) in separate spheres, which bear on them a role of transforming and dissipating agents during a corresponding cycle of angular moment. But common (not balancing) considerations about their origin questions are avoided or at best are transferred to other spheres. Thus, according to some oceanographers, large-scale turnovers here are of storm origin, and in the opinion of meteoro-logists, on the contrary, they are caused in atmosphere by temperature. Therefore, joint balance consideration, of large-scale cyclic motions of the Earth's crust, the ocean and the atmosphere will exclude such interpretations, elucidate their unclear aspects and help to distinguish independent geophysical processes which carry out the above-mentioned transformations and dissipation. Joint considerations will revise sizes, regimes of transfer of angular moment from one sphere to another, and limits of possible rotation. For this purpose it is necessary to divide the balance equation of the Earth's globe angular moment (4) into three components choosing boundary, initial and moment transfer conditions in a convenient manner. But from the nonuniform rotation of the Earth it follows that an angular moment is not constant and, hence, its portions are different from cycle to cycle and for different media. And it means that angular moment distribution according to spheres is performed in an optimum manner, basing on possible reserves of moment in the crust, the ocean and the atmosphere, rates of transformation and dissipation, direct and reverse transfers. But here many questions are not clear yet. According to (Oort, 1983,1988) a question of redistribution of momentum is not studied well. Just in the same way angular moment redistribution in the Earth's crust is not clarified yet (Perstev, Ivanov, 1985).

Here an important fact should be noted. Usually tides, rotation and nutation, as well as cyclic changes of other natures and scales always have higher values and sizes then predicted by the theory (Maksimov, 1970). This is due to the fact that, as a rule, theoretical calculations determine and estimate only average values (temporal or spatial) for isolated excitations. And all this in the case when the observed real effects represent a specific "sum" of the mentioned cyclic changes, and of their separate stages, say results of attribution of maximal. If a certain number of effects are coinciding in space and time, i.e. are arranged spatiality in the specific order say, in one line during a certain time interval and in the results their values are either added or subtracted, then they exceed a simple sum of averages. In other words, they represent a specific superposition of resonance and synchronisation. Hence, the ocean and other sphere counteract not to the average value of this or that excitation but to superposition of their specific stages.

From here it follows that external excitations are not "continuous" but are of pulse nature. For this reason it seems expedient to use a momentum balance equation for low frequency excitations. Thus, a momentum balance that describes forces determining deformation and drift of the Earth crust, is:

$$\mathbf{mD_tU = -mfKU + F - mg\nabla H + \tau_u + \tau_w} \qquad (6)$$

where $\mathbf{D_t = (\partial/\partial t + U_\nabla)}$ is an individual derivative, **m** is mass on a unit area of the Earth's crust, **K** is a surface normal unit vector, **U** is motion speed of the Earth's crust, **F** is a force caused by the internal strains in the crust, **f** is the Cariolis force, τ_u and τ_w are strains in the ocean and the atmosphere, respectively. But up to now it is unclear how to use the equation (4)-(6) in order to distinguish reactions (solutions) of ocean and other spheres to daily effects.

In conclusions we would like to note that it is undoubtedly unilateral to connect specific features of the ocean different-scale cycles only to large-scale cycles (turnover). For small-scale changes in the ocean (say, storm waves) there is a wide spectrum of regular changes of the atmosphere, for medium-scale- numerous excitations of basins and bottom (gradients of rift zones, low frequency oscillations, interwater shock, etc.). But we should consider large turnovers in the ocean as main causes of origin of different-scale morphological structures and cyclic interactions in the ocean, and the above mentioned exogenic, independent excitation contributions as secondary and intermediate.

7. Cycles Search Methods

The mathematical methods' investigations of the ocean cyclic changes and rhythms are subdivided into two large groups. In the first group stochastic, nondeterministic interpretations of oscillations and statements of questions are preferential. In the second one, on the contrary deterministic statements of oscillations are accepted and probabilistic nature of their origin is neglected. But if deterministic models for such cyclic phenomena as changes of oceanic level, its turnovers and other large-scale changes, which are caused by tidal, rotational, nutational and lower frequency regular changes, justify themselves and predict behaviour of phenomena, then for small-scale and mass phenomena, such as spatial and temporal changes of storm speeds, oceanic currents, storm waves and hydrometeorologic properties of oceanic waters, these individual, deterministic approaches are senseless and interpretation of their probabilistic models is more correct. And therefore the stochasticity should be understood not as a consequence of noise or fluctuations but as a result of oscillating activity, i.e. own complicated dynamics of a system with sufficiently large number of degrees of freedom. It is very important that are the mechanisms determining origin of stochasticity in this system, what are the connections with the parameters of the system and which are characteristics of stable stochastic motions. But we are interested in stochastic models of different-scale cyclic phenomena of the ocean represented as time series of natural data.

A probabilistic model is understood as a particular form of random process record (representation). This allows obtain information about any probable characteristic, taking account of a dependence of the process on external conditions. In such a manner that it is necessary to determine an estimation order of probable characteristics of natural data chosen for consideration in accordance with a mathematical model, and to study properties of these estimates.

Usually a structure of random process is investigated by means of methods of harmonic consideration, considerations of period records, revealing of hidden periodicity and spectral expansion. But these methods consider unilaterally closely connected two fundamental features of natural phenomena-recurrence and changes (stochasticity) of their values. Taking this fact into account, some researchers suggest to describe rhythms by means of periodically correlated random functions. Thus, according to Dragan and others, a random function is called periodically correlated one, if its mathematical expectation and correlation function are periodical in 't' with a period T, that is the following relation takes place:

$$\mathbf{m_\xi(t) = m_\xi(t+T)}, \qquad \mathbf{k_\xi(t,\tau) = k_\xi(t+T,\tau)}, \tag{1}$$

A model representing rhythms through periodically correlated random functions is:

$$\xi(t) = \sum_k \xi_k(t) \exp\left(i k \frac{2\pi}{T} t\right) \tag{2}$$

where ξ_k **(t)** is a stationary random process.

Thus a relation of periodically correlated random functions, which characteristics are periodical functions, serves as a model for simple rhythms. Such an approach allows consider within the limits of one mathematical model stochasticity and recurrence of phenomena simultaneously. Then, taking account of well-known phenomena of superposition and interaction of two and more rhythms, the abovementioned relation of random functions is widened and simplest mathematical models of such phenomena are proposed. Thus, according to Dragan a random process is called as a double periodically correlated one if it is represents in the form:

$$\xi(t) = \sum_{kj} \xi_{kj}(t)\exp(i\,\omega_{kj} t) \tag{3}$$

where $\xi(t) = (\xi_{kj}(t))$ in a stationary matrix process. A mathematical expectation and a correlation function of such phenomena are nearly random functions with the following Fourier coefficients

$$m(t) = \sum m_{ln} \exp(i \lambda_{ln} t)$$
$$K(t,\tau) = \sum_{ln} k_{ln}(\tau) \exp(i \lambda_{ln} t). \tag{4}$$

Thus, the class of periodically correlated random functions allows to characterise a wide range of data, such as a mathematical expectation **m(t)**, a dispersion **D(f)**, a correlation function **K(t,τ)**, a spectral density **S(t,ω)**, as well as coefficients of function expansion into Fourier series; $\mathbf{m_k}$, $\mathbf{K_k(\tau)}$, $\mathbf{S_k(\omega)}$. In each particular case, depending on the form of probable characteristics and parameters, a representation of a probable model of the process is revised. Probabilistic representations of rhythms' models have a certain cognition value at the present stage, but they do not take account of two fundamental problems. First, a set of rhythms is subjected to a cer-tain variation principle and has a subordinated structure, and second, which is much more impor-tant, the mechanisms of rhythms origin are neglected completely. To reveal them is important because it will allow to predict and to use the behaviour of rhythms in a way convenient to us.

Deterministic models of oceanic cyclic changes are numerous and diversified. They widely and deeply consider structures and interactions of small- and meso-scale, synoptic, geostrophic, seasonal, large- and other scale changes of the ocean, various problems of origin and stability of dynamics and energy content. The proposed models mostly reflect on, and predict correctly the mentioned cyclic changes and their specifies. Of special interest are the mathematical models of large-scale changes, which are based on the equation common for all changes. Here, as a rule, Bussineskian and hydrostatic approaches are used. Let us discuss one of the most propagated versions (Brayn, Sarmiento 1988). Suppose **U** is horizontal component of a velocity, Δ is horizontal operator of a gradient, **f** is the Coriolis parameter, **P** is a pressure, **S** is a density, ρ is a characteristic density, **w** is a vertical velocity, **k** is a unit vertical vector, and $\mathbf{d_t}$ and σ_t are individual and local derivatives with the respect to time:

$$\mathbf{d_tU+f_kU+\Delta(\rho/\rho_0) = F} \qquad (5)$$

$$\mathbf{\rho g + d_t\, P = 0} \qquad (6)$$

$$\mathbf{U + d_t\, w = 0} \qquad (7)$$

$$\rho = \rho_0\,\mathbf{(1-\alpha\theta)} \qquad (8)$$

$$\mathbf{d_t\theta = Q}\ , \qquad (9)$$

Eq. (5) is an equation of motion, Eq. (6) is a hydrostatic equation, Eq. (7) is an equation of continuity, Eq. (8) is an equation of state, which connects density with temperature, taking account of the effect of fluidity caused by heat and salinity, and the parameter of x is called a coefficient of volume expansion. **F** and **Q** terms describe parametric represen-tations of small-scale effects in the equation of momentum and heat. The following simple parametrization of vertical transfer is assumed:

$$\mathbf{F = A_{\mu H}\,\nabla^2\,U + A_{\mu V}\,d_zd_zU} \qquad (10)$$

$$\mathbf{Q = A_{HN}\,\nabla^2\,\theta + d_z\,(A_{HV}d_t\theta)} \qquad (11)$$

Where

$$A_{HV} = \begin{cases} \infty, & \partial_z\,\rho \geq 0 \\ \text{const}, & \partial_z\,\rho \leq 0 \end{cases} \qquad (12)$$

Eq. (12) represents a simple parametrization of convection. The coefficient of vertical diffusion has a constant value for a stable stratification, and in the case of instability it becomes infinite. Such parametrization provides existence of stable stratification in a model. The following conditions are assumed for the upper boundary:

$$W = 0 \tag{13}$$

$$A_{H}\, d_{z}\, \theta = \mu\, \Delta z\, (\theta^{*} - \theta) \tag{14}$$

$$A_{\mu\gamma}\, d_{z}\, U = \tau^{*} \tag{15}$$

The condition of (13), so called a rigid roof, efficiently separates (filters) external gravitational waves. Eqs.(14) and (15) describe flows of momentum and fluidity on surface. A quantity μ^{-1} in Eq.(15) has dimension of time, Δz is a variable thickness of stratum. The boundary conditions are a function of geographical latitude and are written in the following form:

$$\theta^{*} = \Delta' \Theta\, (1 - 1{,}13\, \varphi) + \theta \tag{16}$$

$$\tau_{\gamma} = \tau^{*}_{0} \begin{cases} 1{,}1\cos(4.77 - 5.46\varphi) + 0.5, & \varphi \geq 0.175 \\ 0.2 - 3.2\varphi, & \varphi \leq 0.175 \end{cases} \tag{17}$$

An integration domain covers two meridians separated by 60°. In the equator the symmetry conditions are assumed. A polar wall is located at the latitude of 60^0, oceanic walls and bottom are considered as heat-insulators. Side boundary possesses wettability,and the bottom is imposed with apparatus condition.

A set of boundary and initial conditions (13-17) and the conditions of stratification (10-12) unambiguously determine a to-be-sought function: distribution and spatial behaviour of horizontal component of velocity **U**. But, on the contrary to statements of technical and artificial problems, a set of these conditions in natural systems is not given forever and is changing continuously from cycle to cycle (in the sphere of biorhythms these changes amount up to 35 to 45 percent of their values).

Hence, a distribution and behaviour of to-be-sought function are changing from cycle to cycle. And it results in the fact that values of **U** are repeated with non-uniform periodicity in time. This phenomenon is characteristic for every hydrometeorologic parameter. In all the cases the analytical models of (11-23) types played and apparently will play a cornerstone role in the problems of stating of new versions. In many cases the experiments carried out on these models help to choose further investigation directions, which allow to understand more deeply and accurately dynamics and interactions of the system and to predict them. Models solve not only idealised and dynamically important problems but also are concerned to the problems of direct observations. In many cases consideration of observed facts help to understand not only the way the model is working but also to clarify the way the ocean is working, what are functions of different scale processes and structures, and mechanism of their origin.

The obstacles in development of models predicting oceanic turnovers are and representative absence of observed data and low resolving power of models. A resolving power includes details of basins and bottom relief in a sense that it should correctly reproduce specific features of turnovers related to them. Good working of the model depends on correct and suitable choose of boundary and initial conditions. Finally it is necessary to find out a mechanism of mixing, i.e. to parametrize vertical mixing process correctly. Conditions chosen for model leave their direct impact on obtained results and this fact should be recognized confidently, i.e. proposed model should not be considered as the real ocean and set off against it. At the best such models can only roughly represent real systems and therefore should be only considered as one of the methods to understand the ocean working. On the other hand, neglecting one or another condition in the model simplify the model to a degree that important physical mechanisms, which are causes of motions and interactions in question, are vanished.

Development of analytical methods for study of the ocean cyclic change depends on cons-truction of models with high resolving power. Thus, if results of two- and three-level models determine a vertical resolving power of a model, i.e. with vertical structures medium and whirl fields, relatively well, then in the case of eight levels vertical structures of processes are expressed much more correctly reliably in the solutions (Brayn,Sarmiento 1988). But improvement of resolving power (vertical and horizontal) should be directed not only to improvement of solution accuracy or sensitivity, but also to inclusion of different-scale changes and interactions.

Thus, analytical models are a reliable basis for consideration of cyclic changes and turnovers of the ocean. The only condition is that they should be constructed in accordance with the certain principle: to try to reproduce changes of different characteristic ranges and to match them with each other. At first models concerning respond of the ocean to low frequency forces such as tides, rotation, nutation forces, etc, which involve the whole three dimensional spaces of the ocean, should be developed. Then these models including stationary models explaining conservative characteristics of the ocean represent a basis for development of models reproducing synoptic changes. Then it is necessary to construct models reproducing changes of meso-scale taking the variables of the abovementioned models as conservative characteristics or parameters. And finally taking the meso-scale variable as slowly changing parameters, it is necessary to construct models reproducing small-scale changes.

Such simulation will allow to revise the whole structure of the ocean relaxation: succession of hydrophysic processes of particular frequencies, which are forming and vanishing every time in each cycle and restore balance within the limits of this scale. It will allow to trace different-scale processes and changes using completely different sets of boundary and initial data and verifying an entire process of dissipation. Hence the main axis of the ocean cycle simulation should be classification of this conservative characteristics, models providing their stationary distributions, which should represent morphological structures (slowly changing variables) of the scale under observation for forming rapidly changing variables, on the one hand, and should be dissipative forms and motions for lower frequency changes, on the other haund. And it is natural that all they should be constructed in accordance with the autosimulation principle.

At present, models reproducing cyclic changes of various scales in the ocean are different only in their resolving power, by a degree of vertical and horizontal parametrization. Resolving power in vertical direction includes processes of turbulent mixing, and that in horizontal direction - instabilities and specific features of turnovers. These models are of a great cognition value, but they do not consider specially problems of interactions between them and energy emission. But the fundamental problem of cyclic changes is particular laws of their direct and reverse interac-tions where every characteristic scale has its own contribution to division of works. Taking account of this fact, it is necessary to pay a special attention to forming of the ocean responds (relaxation) to different scale excitations. But the problem of the ocean relaxation, i.e. time interval instability elimination, is nearly neglected. According to Zimov a time interval for adaptation of the ocean upper layer to a storm is decreasing with the decrease of latitude. It means that a respond of the Ekman's layer to an excitation (storm) is formed faster in low latitudes than in higher ones. It is confirmed by the fact that the known Somalia current was formed in equatorial region within only few weeks (and hence, it confirms its stormy origin and not its balanced state) and that it took years to exile a little the Gulf Stream in medium latitudes. The same short relaxation (high adaptation) and high rates of heat distribution also explain El-Nino current in the Pacific Ocean and seasonal currents in the Atlantic and Indian Oceans. If this phenomenon would be confirmed finally, then polar motions of the ocean may be cause for adaptation of low frequency excitations, which is very important. For example, it may be a serious basis confirming rotation nature of origin of gigantic southern circumpolar current or effect of low frequency about which Stomel said in 1961 that it can not has belt a origin, here their contri-butions have other energy sources.

CHAPTER V

ATMOSPHERIC CYCLES

1. Cycles Nature

Atmospheric cycles occupy a special place in study of cycles. It is due to the fact that here all "hidden" aspects of cycles are displayed evidently. It concerns to their forming conditions, behaviour, significance, spatial-and-temporal classification, energy transfor-mation, evolution and methods of investigation. Quick and correct checking of atmospheric cycles' regularities and their perceptibility make them a specific "source" for clarification and under-standing of similar properties of cycles of other natures. Physical nature of cycles often confuses and impedes to find similarities which are bases for them and which hardly are the results of accidents, and manifests, as physicists like to say, a unity of the Nature.

Atmospheric cycles are air masses turnovers of different scales. This class of phenomena also includes regular changes in temperature (**T**), pressure (**P**), density (**D**), humidity and cloudiness of the atmosphere. Thus they are both general and regional and local regular changes of **V, T, P, D** and other meteorological parameters' distribution in the atmosphere. Nonmeteorologic parameters of the atmosphere such as spatial-and-temporal changes of electric, magnetic and other fields, i.e. their perturbations, deviations and so on, which will not be considered here, may also be classified as cyclic changes.

Problems of forming, preservation and elimination of spatial-and temporal characteristics of atmospheric cycles are directly related to distributions of the above-mentioned parameters and laws of these distributions deformations. The basic features of these cycles in the fact are that they cannot be considered independently on and separately of each other. They are closely connected and complete each other. As Lorenz had noted, motions observed in the atmosphere are impossible to explain without taking account of observation of existing pressure, humidity and temperature fields. The observations of these fields will also be incomplete without observations of wind fields. Despite the circumstances, distributions of the above-mentioned parameters, their general, regional and local changes are studied and observed separatedly from each other. The other specific feature of these cycles is the fact that they are formed by both convection and advection motions (when the properties of air - **T, P** and **D** - are moving together with air masses in the horizontal direction). But what are the forces causing these motions? Let us briefly describe the existing concepts and viewpoints concerning **V, T, P** and **D** fields.

2. Generation's Causes

There are two basic versions concerning origin, preservation and elimination of permanent motions of current-scale air masses in the atmosphere. According to the first version the specific features of atmospheric turnovers (in both meridian and latitudinal directions and whirl motions of different scale, i.e. linear and non-linear motions) have thermal origin and causes. Therefore the mentioned systems of velocities are caused by non-uniform distribution of the Sun thermal energy in the Earth surface. According to the second version, the atmosphere structure, vertical and horizontal permanent and regular motions in it, as well as whirls phenomena of various sizes have dynamically or mechanical origin and cause. In other words, they are related to the specific features of the globe rotation and water rotation, and particularly, are consequences of rotation moment transfer. Let's discuss these problems in more details.

It is known that in the equatorial and tropical regions an averaged annual thermal balance (and not only annual) is positive, and in the polar region, on the contrary, it is negative. The atmosphere is heated in low latitudes, and is cooled in higher latitudes. Since the density of warm air is lesser than that of cool one, then a pressure gradient is formed between the equator and the poles, this gradient, under the influence of gravity, gives rise to a convective motion in meridian planes. Air, under the effect of Archimed's force, trends to move in such way that pressure gradient is eliminated. But since the sources of heating and cooling exist permanently, then this gradient also exists and works permanently, and hence, convective motions are present and preserved always. On the other hand, the Earth rotation speed affects these meridian motions. Due to this effect of rotation a part of the atmosphere, which is moving to the north with a velocity **u** is accelerated with the Coriolis acceleration $\mathbf{u} \times 2\omega$. Under the influence of this force equatorial air mass flowing to the pole turns to the east in a intermediate zone, and forms a new regular motion-eastern winds in the northern hemisphere, and regular eastern winds in the southern hemisphere.

Thus, a part of the atmosphere mass, which is formed in low geographical latitudes, constantly trends to higher latitudes (up to ± 30°), forming so called Gadley turnovers, and another part in the opposite of the former, is formed in cold zone, constantly penetrates to the south, to intermediate zone, and forms the other turnovers, which are called Frenel's ones. Turnovers running in the Gadley's and Fresnel's cells have the same oscillation period, but different energetic powers. Thus, the Gadley's cell transport to the latitudes of ±30° a mass of 230.10^6 t/sec in winter and 175.10^6 t/sec in summer, and the Fresnel's cell transports to the regions of intermediate zone masses of 30.10^6 t/sec and 17.10^6 t/sec, respectively. Thus, due to equalisation of the atmosphere pressure distribution gradient the permanent and regular motions and displacement in mentioned direction from the equator to the pole are formed. This equalisation inevitably causes equalisation of the meteorological parameters **T, V, P,** and **D,** as well as energy and momentum of the atmosphere. As Palmen was noted, a significance of turnovers in the Gadley's cells is rather transformation of the Earth rotation moment into relative moment of the momentum, than equalisation of motion momentum, its efficient transfer or thermal gradient. If the necessary horizontal transfers of the mentioned parameters to the intermediate zone are carried out through the Gadley's cells, then these transfers to higher latitudes, according to some investigations, are carried out mainly at large-scale whirl level. In other words, gradients of **V, T, P, D** and other meteorological parameters are equalised by means of large-scale meridian turnovers and whirl processes, particularly, motion momentum is transferred in meridian and vertical directions. Then, it is found that at north (south) latitude of 20° the atmospheric large turnover and whirl currents are of such an order that they are flexibly adapted to each other and form a powerful cascade of turnovers. According to investigations, existence of high pressure's zone in tropical belt, as well as of the polar cyclonic zones, causes in the lower part of the troposphere stable eastern winds, trade winds, in the equatorial-tropical belt and dominating western winds in polar regions. A separate system of atmospheric regular and stable motions is belt (in latitudinal direction) currents, some of which being propagated for long distances, form sometimes a continuous current around the Earth, etc. This point of view generally fails to explain causes of different scale turnover's origin, and a united approach is absent. Moreover, transfers of thermal gradient and momentum related to it remain vague and unclear.

According to the second version of atmospheric turnovers, an explanation of any turnover is based on the concept of energy and momentum, and a momentum is transferred by force moment only. Since the Earth rotation is existing always then the exchange of its momentum in meridian and vertical directions is also existing always. But, what is the mechanism of transfer and com-pensation of rotation momentum, and what are the consequences of changes in the Earth rotation speed on the crust (and, particularly, atmosphere)? Increase of the Earth globe speed (of the potential of deformation forces) results in squeezeness and contraction of it as an ellipsoid. The latter is performed non-uniformly and so called seven parallel zones of inflection and turn are formed in the crust, which determine joint deformations of the revolving ellipsoid. These parallels are 0°, ±35°, ±62° and ±90°. According to calculations, in the case of increase of squeezeness (increase of rotation speed) the outer surface of the Earth globe is contracted, moving away from the poles and the equator to the parallels of ±35°, or in the opposite case it moves from the parallels of ±35° to the equator and the poles in both hemispheres. Thus the inflection zones are displaced to the lower latitudes in the case of rotation speeding up and to the higher latitudes in the case of rotation slowing down. Comparison of these displacements with the axis of high pressure tropical zone shows that these two phenomena are obviously coordinated, i.e. oscillation of inflection parallels positions and displacements of high pressure tropical zone axis position are performed simultaneously and in parallel. The mentioned parallelness of the positions is observed during both month and year. Thus, in the period of the Earth rotation speeding up, which falls on winter (summer) months in the northern (southern) hemisphere, the high pressure tropical zone occupies the lowest latitudes, and in the period of the rotation slowing down in summer (winter) months it moves to higher latitudes. The maximum pressures of high pressure's tropical zones in both hemispheres are reached in July-August and the minimum pressures in February-March, what undoubtedly is related to the Earth rotation speed. The high velocity of the Earth motion in the orbit is reached in August, and the lowest one in February. In other words, a seasonal prevalence of pressure oscillations are related to the Earth motion non-uniformity. The above-mentioned coincidence is not only temporal but also spatial. Thus, pressure distribution in the lower layers of troposphere is of latitudinal, belt nature, it has two maxim: in the region of ±35° parallels and at the pole; and two minima: at ±62° parallels and at the equator. It is noteworthy that such belt-like pattern is also observed in the Earth's crust. Elevations and planes in morphological structures are located in the direction of the same parallels. Moreover, belt structures are observed not only for meteorological parameters, particularly, regularities of pressure distribution, but also for geophysical parameters distributions: magnetic and electric fields and not only in the lower part of atmosphere - the troposphere, i.e. this distribution is sensed everywhere, what evidences on general (planetary) nature of affecting forces.

Hence, during the changes of the Earth rotation speed the accelerations caused by deformation forces and affecting paralleled plane in latitudinal direction, as well as on meridian plane in the directions of tangential and normal components, are arisen. The atmosphere, being the most mobile component, immediately responds and repeats these cyclic changes in its structure, trying to dissipate effect of changed motion momentum during the cycle.

Then numerous existing facts show that increases and decreases of northern and southern centres of atmospheric high pressure (and apparently not only of pressure) are simultaneous and parallel. First, it evidences that the atmosphere behaves itself as a whole with respect to the deformation forces during the Earth rotation and its responds are uniform in the sense that turnovers and motions of all orders are inevitably born, caused by and adapted to the momentum transferred to the atmos-phere and structures are reconstructed in such a way that they trends to discharge and neutralise

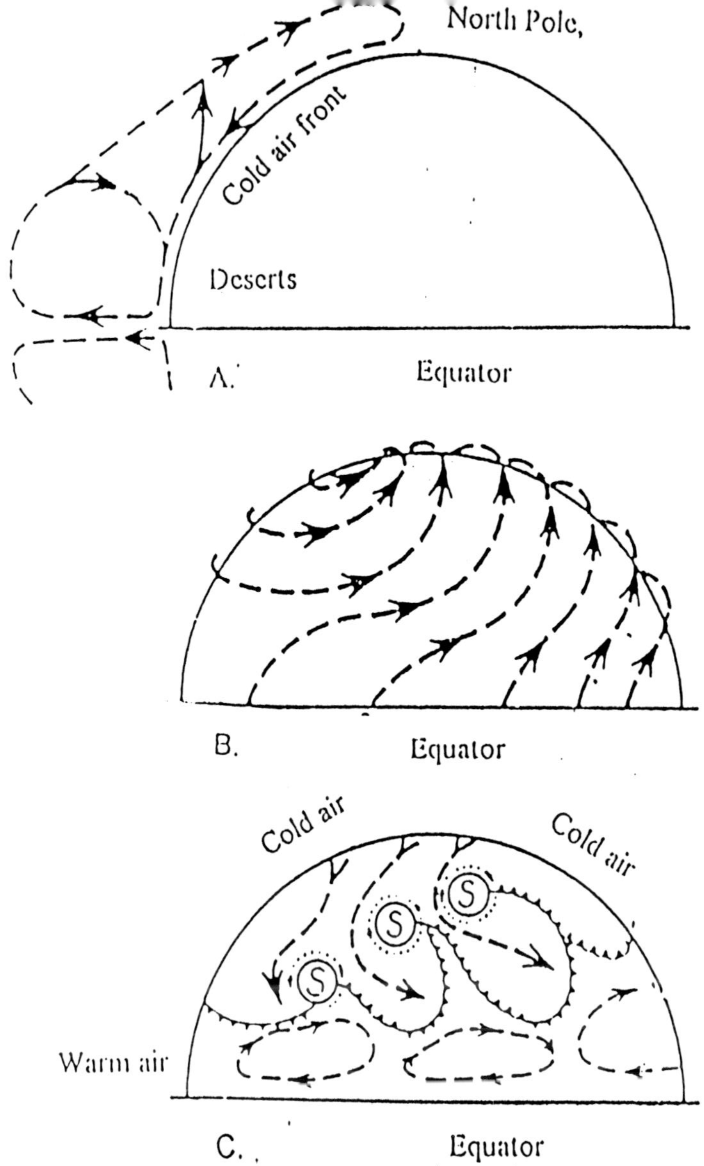

Tropical air is elevated at the equator and is descended over deserts at medium latitudes and poles

Tropical air is elevated at the equator, approaches the pole helical at high lablitudes, is descending here

Whirls originate along borders polar fronts -division between cold and warm air. In the result stormily winds polar air breaks fronts and directs to south, to equator

Fig..16 Energy system of the atmosphere. push out warm air in tropical, loons are formed in polar fronts (after Odoum G. Odoum E).
A. General alternation of turnover.
B. Winds in the upper layers of the atmosphere
C. Winds on the Earth surface

new transferred motion momentum. Second, it shows that the order of magnitude of thermal gradient between the equator and the pole is smaller than that of the rotation momentum transferred to the atmosphere.

The regularity of high pressure distribution allows some researchers to look for hereditary connections between them and **1.** Eastern winds within the belt of ±35° of the ocean and atmosphere equator; **2.** Dominating western winds and jet currents in intermediate latitudes from ±35° to ±62°, and **3.** Dominating eastern winds in polar regions in the belts from ±62° to ±90°. Thus, in the opinion of some researchers, the changes of deformation forces potential and rotation regime cause by the increase of the atmosphere activity in the regions of inflection parallels 0, ±35°, ±62°, ±90°. But what is the mechanism particularly responsible for transfer of rotation momentum to these turnovers of the atmosphere remains weakly substantiated. Nevertheless, despite the detailed investigations of separate turnovers, there is not their complete consideration and the relation with thermal gradient is not clarified yet. It should be suggested origin that changes of the Earth rotation, thermal gradient, tidal effects, volcanic and other cycles contribute to forming and preservation of different-scale turnovers of the atmosphere. In the course of time superposition of all these highest and lowest values often complicates rhythmic behaviour of different-scale atmospheric turnovers.

3. Cycles' Classification

What domain of frequencies do they occupy? If a unit of one oscillation per hour is chosen as a measuring unit, then atmospheric cycles, as literature data show, may be included in the following domain: from 10^{-9} to 10^{4}. This domain includes regular atmospheric changes lasting from parts of second up to hundred thousands and more. As in all other cases here also researchers for sake of simplicity divide this domain of frequencies and events into parts and groups. Thus, some of researchers conditionally subdivide this domain into 10 groups. Each group of oscillations has its own name and is determined by oscillation which energy spectrum dominates over other members of the group. These groups are the following:

1. (10^4 -10^2). This frequency domain includes oscillations lasting from parts of second to minute. This domain of frequencies is called micro meteorological. Here the dominant oscillation is turbulence of small scale. Its energy spectrum in the near-ground layers has duration of maximum one minute in the vicinity of $\tau < 1$ oscillations, which is corresponding to a non-uniformity of horizontal turbulence of $L = U\tau_{max} = 600$ m scale. Causes of these oscil-lations origin's are several: a) turbulence; b) sound; and c) gravitational waves of short periods, τ being equal to 300 sec for them.

2. Oscillation from **(10^1 - 10^0)** minutes up to one hour. They are called mesometeorologic, the dominant oscillations are gravitational waves and waves caused by atmospheric convections. The characteristic scale of these oscillations is **$L = u\tau = 10$ km.**

3. 10^{-1}. Rhythms of this frequency are cyclic changes of periods almost close to it. The causes of origin of these oscillations are two: thermal radiation of the Sun and tidal waves. Daily oscillations of atmospheric pressure take place mainly in the form of atmospheric tides, related in the same manner with the daily oscillations of other meteorological parameters, as well as humidity, cloudiness and winds.

4. Cycle of 10^{-2} frequency has duration of more than 24 hours and up to several days. These oscillations are called synoptic. The changes of this order are studied relatively deeply and comprehensively. The maximum of energy spectrum of these oscillations falls at 4 days. The main forces giving rise to such oscillations are cyclones and anticyclones of intermediate zone. According to some researchers (S.S.Monin et al) due to synoptic motion with high Reynolds number **Re** = **UL** / ν almost horizontal hydrodynamic instability takes place in higher part of the synoptic domain from large-scale motions of spectral energy cascade transfers to small-scale motions. And in lower frequency domain; in the opinion of other researchers, the reverse transfers take place: energy from synoptic motions is transferred to motions of larger scales. Synoptic motions of all scales, because of vertical non-uniformity of wind fields and hydrodynamic imbalance, give rise to microturbulence immediately and constantly. In other words, micro-turbulence is a dissipate factor for synoptic motions, and for this reason it is characterised by **ν**, efficient turbulence coefficient of viscosity.

5. 10^{-3} Cycles of this frequency are called global, their duration is about one month and time intervals amounting its parts. These oscillations are originated and caused by latitudinal turnovers of intermediate zone, self-oscillating interactions of the atmosphere and the ocean. The maximum of energy spectrum is equal to two weeks.

6. 10^{-4} Oscillations close to this frequency are called seasonal. Their period is nearly three months and is originated depending on alternation of seasons. More precisely, it depends the changes of the angle made by the plane of the Earth rotation axis and the ecliptic. The oscillations of this order include famous monsoon winds, which change their speed and directions depending on the seasons: in summer they are directed from land to sea, and in winter from sea to land. Annual and half-yearly oscillations of the atmosphere also affect (strengthen or attenuate) seasonal cycles. Thus, half-yearly strengthening connected to oscillation of the Sun radiation in near tropical latitudes. In the near-tropical region the Sun passes the zenith twice a year (at days with equal-nights on the equator). Because of this fact, drastic changes, jumps of atmospheric pressure take place between near-tropical regions of land and the ocean, these changes also affect the atmosphere. Among the effects of annual oscillations the followings may be mentioned: two minima (in winter and summer) and two maxima (in spring and autumn) are observed for the annual behaviour difference of temperature and radiation (insolation) in medium and high latitudes. For these reason meridian temperature gradients reach maximum in medium latitudes in transient seasons, and latitudinal turnovers are increasing significantly.

7. 10^{-5} Oscillations close to this frequency are called inter-yearly. Duration of oscillations of this order, i.e. their periods, are up to 3 to 4 years. Causes of their origin are several: a) large-scale interactions of the atmosphere and the ocean; b) motion nonregularity; c) oscillations of the Sun radiation regime; etc. The most significant oscillations of this order are 26 months regular turnovers, which exist at ±10° latitudes in the equatorial stratosphere. It is not clear what is the function of this cycle in the atmosphere and what is its feeding source. The oscillations of the remaining 3 orders are studied much more poorly.

8. 10^{-6}. They are so called intracentury turnovers causes of their origin being unknown.

9. 10^{-7} . Oscillations of this order are called intercentury turnovers. It is suggested that the causes of these oscillations may be selfoscillations of continental glaciers and the ocean deep layers.

10. (10^{8}-10^{9}). They are time intervals of glaciations. In the opinion of researchers, the oscillations of such period (100 to 200 thousand years) are connected to changes of the Earth's orbit, in particular, of its ecliptic (**ε**), eccentricity (**e**) and semi-major axis (**a**).

There also exist other classifications of the atmosphere turnovers. But they are of little cognition value since they are of particular nature. Thus, the troposphere large turbulent motion spectrum is conditionally subdivided into three parts, and only motions in latitudes of 45°, which naturally gives a little representation of problems concerning origin of different-scale motions, their functions and revealing of cause and effect connections

4. Atmospheric Cycles' Structures

As it was mentioned the atmosphere cyclic changes of various scales are closely connected to regularities of its meteorological parameters distributions: pressure (**P**), temperature (**T**), density (**D**), velocity (**V**) and humidity. These parameters jointly form and compose the internal structure of the atmosphere. The atmosphere, being the most mobile of the Earth's shells, has the lowest values of density and viscosity, and it seems that **V,T,D,P** and other parameters on its surface should be balanced and distributed approximately uniformly. But, as investigations show, these parameters on the globe surface and in the direction of its radius have definite distributions, are preserved for a long time and do not vanish, which evidences on constant inflow of external energy. According to a number of investigations, the distributions of **P,T,D,V** parameters have a cell structure, that is, they have high values and concentrations only in particular directions. Interconnections of these directions form a specific net. The parameters of **P,T,D,V** and other meteorological parameters have opposite values inside the cells and on their boundaries. Thus, a temperature inside the cells is much higher than on the sites of their separation, and on the contrary to that, values of **P,D,V** parameters are of opposite location, i.e. they are higher on the sites of separations (fronts) than in internal areas of the cells. However, non-uniformity of the Earth's crust, its surface and diversity of excitation sources somewhat masks the cellular structure of the atmosphere. Just this distribution of meteorological parameters forms and causes so called jet currents and motions of different scales in the atmosphere, these current and motions joining form turnovers. The roots of jet currents are not studied comprehensively and an integral pattern of their forming and preservation is missing. Origin and forming of separation lines and fronts, according to Palmen, are very complex phenomena. According to some researchers, they are connected to deformation and transfer fields caused by the Earth rotation. Observations show that formation and movements of fronts take place in oppositions of the surface and relief. Thus, contour of the Earth surface and relief affects forming of fronts. And generally, frontal and separation lines are formed and preserved in all these cases when factors assisting in this are different and opposite. Palmen thinks that front origin takes place due to existence of certain features say, long-term concentration of temperature gradient. It is obvious that such an explanation may elucidate to a certain degree complicated questions of separation lines origin in the oceans, the Earth's crust and other sites of space. Actually, in all cases motion or frontal lines represent separation or division sites of masses (gaseous, liquid and solid) possessing different physical properties. Significance of different-scale division, independent on medium where they are formed and preserved, should be related to an equilibrium state of a system. Considering a role of frontal lines in atmosphere, Rosby expresses such a thought. According to him, a division of the tropi cal high pressure's zone into separated cells represents a necessary adaptation of the atmosphere, which provides a transfer of absolute momentum from low latitudes to higher ones.

Thus, despite an active mobility of the atmosphere, it possesses a particular structure that represents by itself stratification in the directions of latitudes and meridians, as well as of the Earth radius. These directions are sites of highest or lowest values of the meteorological parameters (**P,T,D,V**), and display an emphasised stability in the course of time.

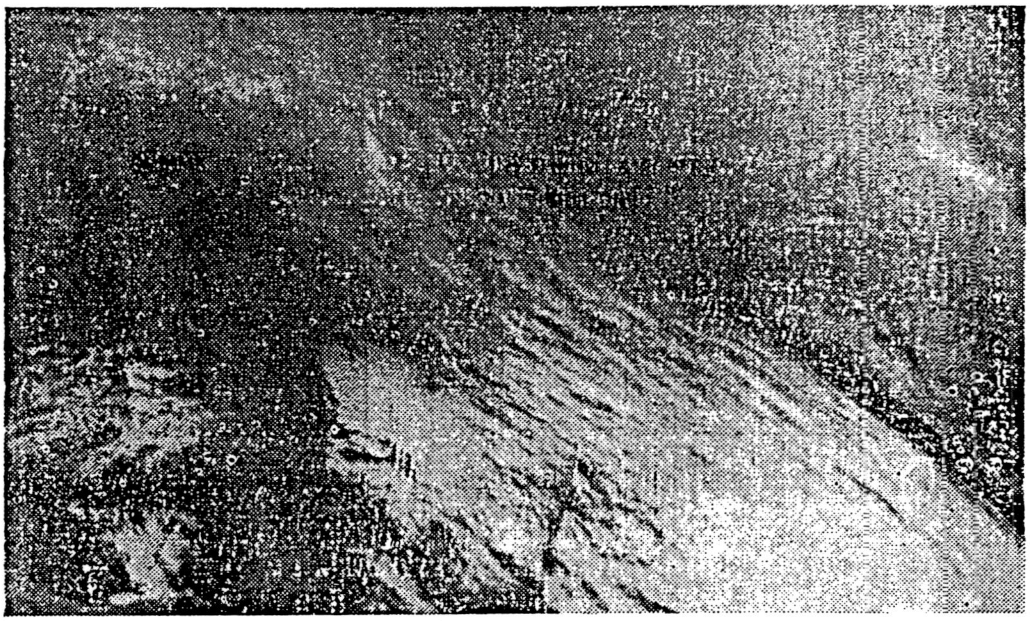

Fig. 17 Powerful frontal, boundary air current in the medium layers of the atmosphere (South-Eastern Europe, Middle East, 1988).

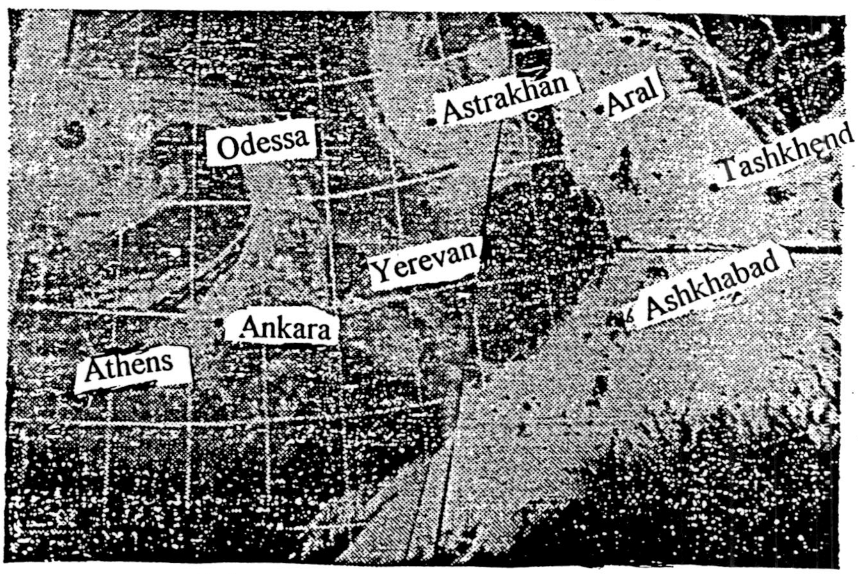

Fig. 18 Powerful cyclones, whirls in the medium layers of the atmosphere (at the same place and time as in the Fig.17).

As it was already mentioned, separation lines and fronts also exist in the Earth's crust and its ocean, which being interlaced with each other from specific nets and are elements of these media structures. According to some researchers, directions of separation and frontal lines of these three media nearly coincide and their net structures apparently replicate each other, forming their continuations. This parallelism and similarly in structures could not be accidental. It shows that, except for local and regional forces, there exist planetary forces, the structures being formed and preserved under permanent influence of these forces. The lithosphere (hydrosphere, atmosphere) responds to permanently acting effects of medium (in this case, changes in the Earth rotation angular speed) by restructuring and redistribution establishing a new equilibrium. Structural similarities allow Rosby to say that any theory of front and jet current formation and development must be applicable, according to its basic physical provisions, to atmospheric and oceanic systems. Apparently, it must be also applicable to the Earth's crust and structures of other scales, since a similarity between current systems, being related to each other indirectly and independent completely by their origin mechanism follows from the most general dynamic principles.

5. Cycles' Significance

The considered turnovers have different functions and significance in the atmosphere, which are determined by local, spatial, general and interterrestial nature of causing forces. Thus, in the case of very-long duration cycles the atmosphere is not distinguished from the Earth's crust oceanic and continental masses and is an integral part of these masses. In the case of long cycles, which are of general (planetary) nature, the atmosphere itself represents a unity, and it's responds unity is directed to neutralise and dissipate a motion momentum transferred to it. Medium-size spatial turnovers are directed to neutralise and dissipate a motion momentum transferred to air spatial masses, or otherwise, a moment proportional to momentum. These turnovers also perform a function of balancing of permanently existing meteorological and other quantities' gradients in the atmosphere. Short turnovers are of local nature and perform dissipate function due to opposite sign of viscosity. In other words, because of small turbulence, the insatiable motions of large scale are decreased gradually, and are converted to heat because of viscosity. A negative sign of viscosity, which is classified as very important by Lorenz and Starr, performs the energy transfer in reverse direction, that is from small scales to large ones. A viscosity being of negative sign is not excluded, a question concerns its duration, that is how long is it presented in the geological time axis. Because a positive sign of viscosity should explain, the atmosphere evolution, which here is not as significant and evident as the evolution of lithosphere. Evolution cannot be avoided because this avoidance would mean that questions concerning, it concerns also a system stability, and substitution of these questions by the others.

The mentioned atmospheric turnovers of all orders are characterised by the fact that their trajectories are approximately straight lines. Most of these turnovers, as it was noted earlier, are jet, belt currents, which flows periodically are increased and decreased in the course of time.

Besides straight-line motions, there exist so called nonlinear, helical motions in the atmosphere, which trajectories are not straight lines. These motions include different-scale whirl currents, cyclones and anticyclones. The whirl motions posses a number of specific feature: a) they are not closely connected to any geographical latitude and longitude; b) their life duration is incomparably restricted; c) conditions of formation, preservation and decay and cycle recurrence are not clarified finally; d) a level of their knowledge is lower than that of linear motions. But it is well known that: 1)

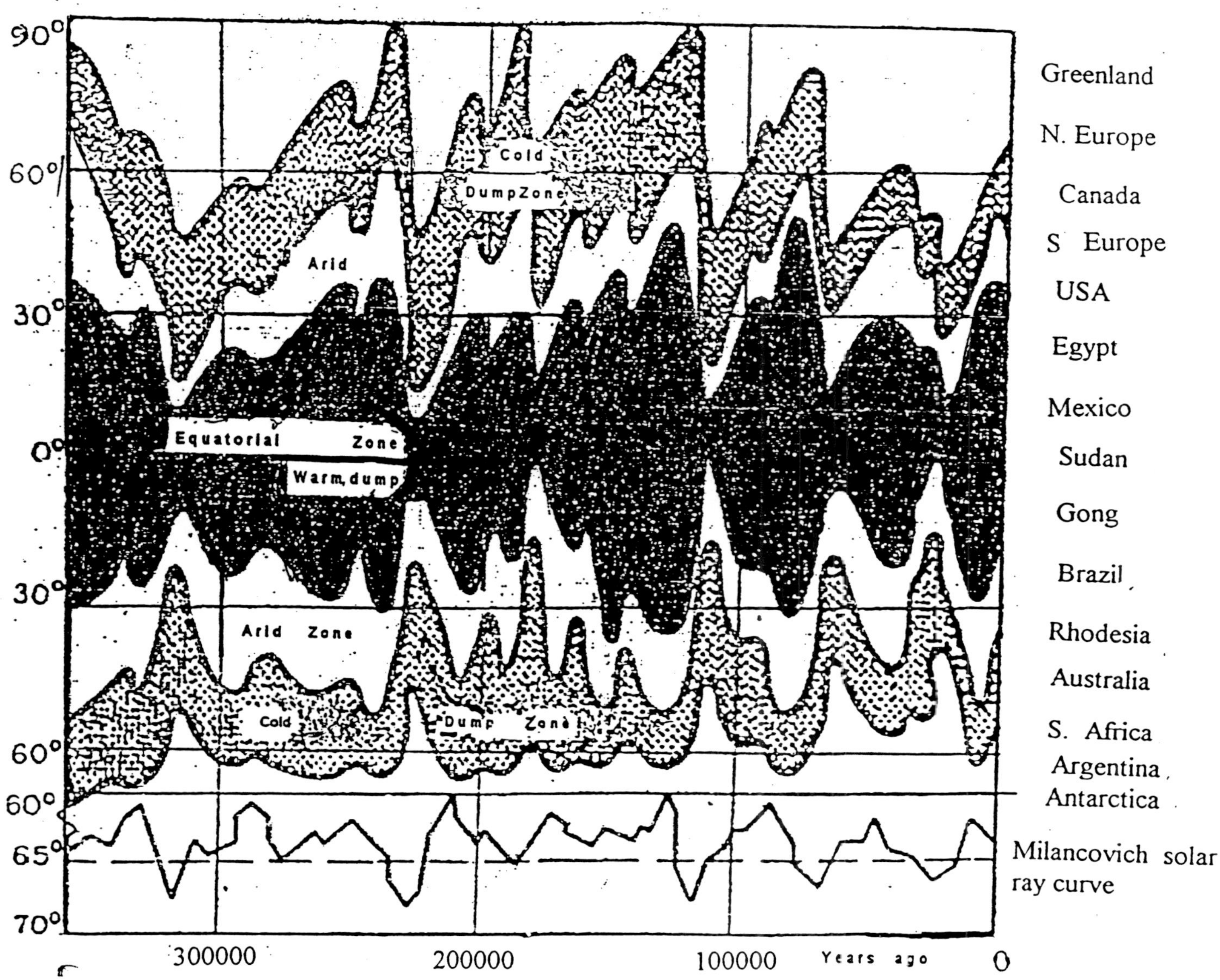

Fig. 19 *Oscillation of the principal Zones of the Earth during the past 3.5 million years Displacements of climatic Zones, especially in the North hemisphere, where 2.000 to 3.000 km (after Faibrige, 1963)*

whirl motions are formed mainly to the north of medium latitudes; 2) motion momentum is quickly and reliably transferred; 3) transfers occur only in horizontal direction and not in vertical one. But, as Lorenz had noted, it is understandable how are the absolute and relative moments of motion momentum transferred to and realised by whirls. According to some assumptions, the cause of whirl motions may be a necessity to transfer an excess of energy or to replenish its deficit within separation frontal zone of the atmosphere. In other words, they may be born within the inhomogeneities of the atmosphere, i.e. there where meteorological parameters have high gradients, which should be recorded. If atmospheric latitudinal turnovers are the mechanisms which realise dissipation and elimination of motion momentum transferred to the atmosphere, then whirl motions represent those mechanisms which balance incompletely compensated gradients (of momentum, thermal, etc.) inspiration zones.

6. Cycles' Search Methods

What is the methods for investigations of atmospheric turnovers? Large-scale turnovers are studied isolatedly and individually, and small-scale ones by statistical methods. For these purpose researchers try to estimate principal processes, which determine energy and matter flows passing through boundaries have considered volume, regularities of their distributions and redistribution. Possible changes, of the atmosphere physical and other parameters' fields, are tried to be estimated in this manner, depending on temporal and spatial regimes of in-flows and out-flows. But revealing and description of conditions of various-scale changes, and evaluation of their relative roles are very difficult tasks. This is the reason why researchers are restricted to qualitative or approximate consideration when explaining causes of physical field changes. In joint and comprehensive (statistical) considerations of the mentioned fields it is assumed to consider their characteristics and properties as random phenomena and to carry out investigations by means of random functions. Thus, the representation of the atmosphere state evolution as a random function allows to state question of this process statistical approximation. Why statistical methods? Because the field of velocity **V** is considered, and it is the same as to consider distributions and dispersions. The latter allows to find every component energy of the field as a certain portion of total dispersion and hence to find a spectral structure and expansion of the field.

The most general and compact description of a random function F(x,t) changes in time and space is its four-dimensional spectrum:

$$G(K_j\omega) = \int R(\xi, \delta)e^{-i(\alpha+\beta)t}d\xi d\delta \qquad (1)$$

where ω is a frequency, **Kj** is a wave number, and **R(ξ, δ)** is a four-dimensional self-correlation function. But it is impossible to give such a four-dimensional description of any physical field. In such cases, usually two-dimensional investigations are carried out, and separate domains of frequencies or wave numbers are considered. As it was noted, the determination of spatial spectrum shows that portion of the kinetic energy, which is given to this scale motion. Therefore the determination and study of the spectrum will help to clarify motions according to their spatial scales. It is obvious, that some domains of the spectrum may be common and be manifested themselves. In considerations of the velocity field, a so called Reynolds's number **Re=Lu/v**, where ν is a characteristic velocity of flow, **L** is an external scale of turbulence (or a scale of energy transferring whirls), and γ is a kinematics viscosity, are of substantial significance. The number **Re** is a dimensionless parameter and shows relation between inertia and viscous forces in flows. If **Re >> 1** then, according to the theory, whirls when formed from large-scale motion are insatiable and when decaying they born whirls of smaller scales, transferring kinetic energy to them. And if this number for smaller whirls is still less than a unity, then they again being insatiable decay and form more smaller-scale whirls, transferring their energy to them and so on. This process of decaying of whirls into smaller-scale ones is continuing down to those scales, where **Re < 1**, that is, down to those limits where, viscous dissipation is already substantial and significant. Thus, turbulence is determined by a kinematics viscosity (ν), speed of kinetic energy turnover (ε) and isotropy or homogeneity of whirls. Those spatial scales where a speed ε of kinetic energy transfer in a stationary state is equal to a speed of its viscous dissipation ν and where whirls become isotropic and homogeneous are called internal scale of turbulence and are determined by the relation $\lambda = (\nu^3\varepsilon)^{1/4}$. Thus, according to the theory, in turbulent flows cascade transfers of kinetic energy from large scales to smaller ones and dissipation manifests itself only beginning from the internal scale of turbulence. In other words, kinetic energy dissipation does not occur beyond the internal scale. But it is not difficult to ascertain that dissipation also occurs beyond the turbulence internal scale, and even more in the vicinity of every scale. In order to be ascertain in this fact let us to clarify the following question: what scales of motion in the atmosphere or in the ocean may be studied isolatedly and individually, and from what scales should they be studied by statistical methods. Usually this limit is connected to time interval subjected to study. Thus, if a period of motions to be studied is several hours, that is natural that even meso-scale turnovers should be studied isolatedly, and if the consideration time is of order of thousand years, then it is senseless to study isolatedly even inter-century turnovers. In such cases it is expedient to have their statistical or averaged characteristics. Then in order to consider an evolution of the given scale turnovers individually, at first it is necessary to have its initial values and data. But as the investigations show, at present small-scale velocity field is recorded by a network which recording points value of horizontal step are apart from each other on a distance greater than the internal turbulence scale $\lambda = (\nu^3\varepsilon)^{1/4}$ by a factor of 10^7 to 10^6 . Therefore it is impossible to register and to study motions with scales smaller than the mesh cell. Such motions should evidently be studied statistically. Thus, those motions are studied statistically which scales are smaller than the recording mesh cells and which initial data are difficult to find. Now, if a dissipation is allowed only for motions with scales smaller than mesh cell, then it is not difficult to see that it should be absent, according to the theory, in motions which are not included because of the mesh cell reduction. If these speculations are to be continued, then further reduction of mesh cells should cause absence of dissipation in motions subsequently excluded from the mesh. Thus, a presence of dissipation is connected with value of mesh cell, which undoubtedly introduce a certain uncertain-ty. Hence, it should be suggested that dispassion exists in turbulent motion of every scale. Exis-tence of dissipation at every stage is seen more unambiguously and surely in nonhydrodynamic systems. It is known that at every sequential stage of ecological and physiological rhythmic changes of various scales about 70 to 80 percent of inflow energy is dissipated and the remaining portion is transferred and used to preserve the given scale rhythm.

7. Energetic Relations

According to the literature data, most of atmospheric cyclic changes are related to motion (velocity) fields. Therefore, concept of energy here is directly connected to a kinetic energy of motion . For this reason investigations are directed to deter-mine kinetic energy sources, its outflow, as well as its transfer to various scale motions. Efficiency of hydrodynamics methods and principles in this problem is recognised widely. But such an approach dose not reflect completely interrelation of the system's (atmosphere) internal potential and kinetic energies, and there is possible transformations. Hence, the methods, which allow reveal these interrelations, are important. Usually this class of methods includes balance relations. Balance's equations help to evaluate internal and potential energies (structure and phase transitions) of a system, necessity of internal motions (kinetic energy), in order re-establish stability according to the changed conditions of medium. Thus, all the internal reformations and transformations (expenditures of kinetic energy) are caused by external excitation and are directed to establish equilibrium with respect to it, and every time on recurrence of excitation these reformations (responds) run in the same course. Unification, regulation and directions of responds in the course of time results in evolution (phase transitions), structural complexities of a system. Usually a growth of internal structure of a natural system, or its organisation levels results in increase of system stability and potential energy expenditures. If in a changing medium a system organisation level is decreasing, then in order to preserve it an additional potential energy is to be applied. If means that a system, independent on its physical nature, and preservation of its structure inevitably require a certain expenditure of potential energy, which it receives from outside. The given considerations are equally applicable to the atmosphere and its structure, where motion momentum and thermal radiation flows are to be considered as a source of potential energy. Basing on the above mentioned facts it may be said that the atmosphere structure and its various scale turnovers posses a particular direction and driving cycle they continuously dissipate transferred energy of motion momentum and thermal radiation flows. A portion of potential energy is spent to preserve the structure and other one is transformed into kinetic energy and is dissipated through various-scale turnovers of the atmosphere. But what are particular mechanisms of the above mentioned transformations is not clear up to now. In this connection it is expedient to remember the question Lorenz stated: what is the way by which a portion of the atmospheres internal energy, caused by the sun radiation energy and transfer of motion momentum, is transformed into kinetic energy? Then, interactions of motion relative momentum, tidal waves, thermal and other gradients are understandable sufficiently. According to Lorenz, transfers and transformations of motion momentum in the atmosphere turnovers are of more complicated nature than it takes place in the classic hydrodynamics. Here the turbulence is related not only to velocities, but also to pressure, density, temperature and other meteorological parameter fields, which can accumulate or condense different portions of energy and transfer and transform them at the required instant.

Generalising all the above mentioned interactions let up try to write to the simplest balance equations of motion momentum of the Earth's crust-ocean-atmosphere system:

$$\frac{dx}{dt} = a_{11}x - a_{12}x - x(a_{13}y + a_{14}z) + a_{15}y + a_{16}z$$

$$\frac{dy}{dt} = a_{21}xy - a_{22}y - a_{23}yz - a_{24}y + a_{25}z + f_1 \qquad (2)$$

$$\frac{dz}{dt} = (a_{31}x + a_{32}y)z - a_{33}z - (a_{34} + a_{35})z + f_2$$

The first equation shows changes of the Earth's crust motion momentum. The term $a_{11}x$ represents a motion momentum of external excitation transferred to the crust, $a_{12}x$ is that portion of energy which is spent to preserve the Earth's crust structure, this term conditionally includes those portion of energy which are spent for origin of elastic, plastic, thermal strains, displacements of separate sites of the Earth's crust and other relatively short processes. The term $a_{13}y + a_{14}z$ represents that potion of energy (motion momentum) which is transferred to the ocean (y) and the atmosphere (z) by means of interactions. Then, the terms $a_{15}y$ and $a_{16}z$ represents energy transferred to the crusts from the ocean and the atmosphere, respectively due to a feedback (negative friction, viscosity), and particularly, motion momentum. The second and third equations are interpreted just in the same way. Here only f_1 and f_2 excitations are added they represent separate sources of energy (say, weak tidal waves). For sake of simplicity in the set of equations (2) it is assumed that motion momentum is transferred through direct connection and feedback instantly, without delay. Just for the same reason it is assumed that changes have the same rate in three chains of the system, and coefficients are constant. The mentioned set of equations represents balance relations of one common excitation. Such set of equations should be also written for other excitations, particularly, for the sun radiation and its gradual transfers. Then, interactions and transformations of motion momentum of thermal radiation flow should be considered. And of course, the coefficients should be determined or evaluated. Among them the most important are the coefficient's a_{i2} **(i=1,2,3)** which determine occurrence of regular changes of various scales. Such a general approach, without doubts, is far from being satisfactory. But inclusion of atmospheric turnovers (and not only of this nature) into a united system is useful because it allows to outline a sphere of their activity and their roles in a system. Then it is necessary to move forward, when they are clarified to some degree: to perform a new division of works proceeding from the former one and so forth. Now, when motion momentum, which is subjected to dissipation, is revised, it is necessary to determine medium (so called boundary) conditions, where the momentum should be realised. These conditions include sphericity of the Earth, its size, rotation speed, locations of oceans and continents, their positions, shapes, surface, relief and oceanic turnovers. These two groups of conditions determine atmospheric turnover unambiguously. But in order to determine their scales and specific features of origin, it is necessary to have a correct understanding of possible transformation and directions of internal, potential and kinetic energies. As it was noted the problem of energetic of various-scale atmospheric turnovers is a very complicated one and is not revealed finally. According to the investigations, when a directions of air motion coincides with a direction of gravity, or is opposite to it (e.g. upward and downward motions), then a potential energy is converted into kinetic and vice versa. Such transformations take place in Gadley's and Frenel's cells. Just in the same way, when an air motion direction coincides with a pressure gradient direction, or is opposite to it, and hence, is normal to isobaric surface, from high pressures to lower ones (or in opposite, from low pressures to higher ones), then internal energy is transformed into kinetic one and vice versa. In both cases the processes are adiabatic and reversible thermodynamically. Then, air motions, directed in opposite to friction force, and heating coursed by it again result in transformation of kinetic energy into internal one, but now this process is irreversible. Taking account of these regularities of energy transformations, as well as the fact that the velocity field comprises belt-like and whirl components, that is

$$\mathbf{dM/\,df = (C_\varphi + I_\varphi)\,[dy - d(C_z + I_z)]} \qquad (3)$$

where $\mathbf{C_\varphi}$ and $\mathbf{C_z}$ are meridian and vertical turnover flows, and $\mathbf{I_\varphi}$, $\mathbf{I_z}$ are corresponding whirl flows, some researchers calculated various-scale turnovers energy. According to these calculations, there exist a closed circuit and turnover of transformations of belt and whirl motion potential (internal) and kinetic energies. According to this concept, a potential energy of belt motion is at first transformed into a potential energy of whirl motions, then it is transformed into a kinetic energy of whirl motions. The latter then is transformed into a kinetic energy of belt turnovers, which then is returned to the initial state. Such closed circuit of energy transformations, apparently, should be considered as self-regulating system in a sense that this system is subjected to structural deformations in dependence on input (magnitude of excitation) and conditions of medium, particularly, changes of parameters ω and **r**. At the first sight, it seems that feedback somewhat masks directness of phenomena occurring in system and general tendency of energy transformations. But it is not so, beause simple calculations show that they are directed, in a sense, that about 80 to 90 percent of imput energy is dissipated

Table 11.

Energy Transfering and Converting Lorenz Cycle with data 1958-1963.
Energy magnitute is 10^8 joul/m^2, transfering velocity measured wt/m^2 (after Oort)

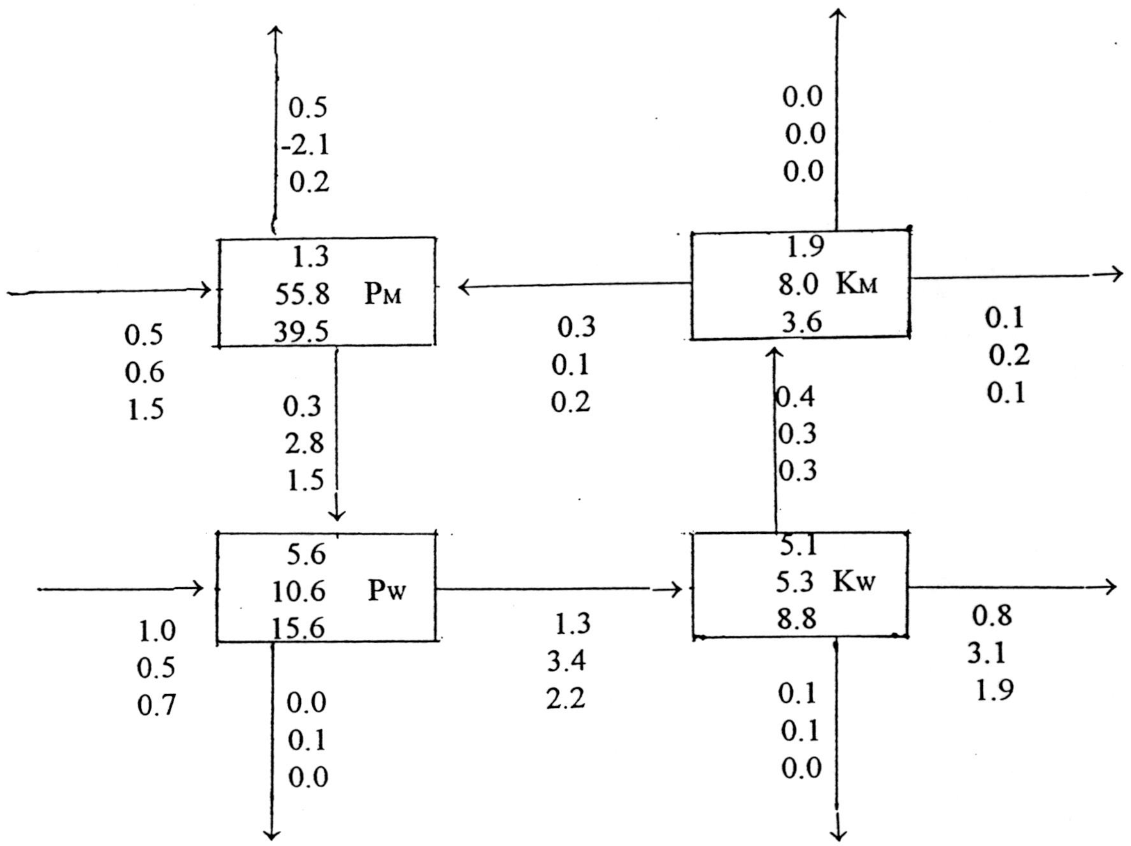

every time within the limits of the given scale turnover and is excluded from the further play. As it was mentioned already, about the same amount of energy is dissipated in biological (81% in ecological, 86% in physiological) systems. What does this similarly hide, it may be only guessed.

Concluding the above mentioned, what may be said about various-scale cyclic changes of the atmosphere in the Earth activity? First of all, it should be noted that the atmosphere, together with the crust and the ocean, is one of consumers of motion momentum. It, being the most mobile among the components and smallest by its mass, instantly responds to every trifle change of motion momentum and is consistently adapting to it. Dissipation take place gradually and subsequently through various-scale regular displacements (cycles) in such a way that on the background of large-scale changes medium-scale changes are formed, and smaller-scale changes are formed from the latter, and so on.

Atmospheric cycles have emphasised spatial-and-temporal subordination that is clearly seen in cycles of all orders. Then, it is observed that they have no constant periods, and can "freely" move in a certain domain, being a mechanism providing stability of a system in the in the changing medium conditions.

CHAPTER V

CYCLES' ORGANIZATION ELEMENTS

The preceding chapters briefly describe the well-known particular cyclic changes observed in various media. Cycles display everywhere, independent a medium, the same organisation: subordination, recurrence and regularity (variation nature). Up to now spatial-and-temporal characteristics of cyclic changes were mainly considered independent on each other and such an approach was caused by the requirements of time. Theirs comprehensive, as continuation one of another, considerations were not paid much attention. But the comprehensive considerations of cycles are of great importance and are useful from several viewpoints. First, the problem of cycle nature, origin and formation is solved relatively easily by such approach. Second, the dependence problem of system parameters on cycle (and not on time, as it is suggested usually) and its evolution are simplified, and then boundary and initial conditions become definite, despite they are changing from cycle to cycle.

What are general features of simplified equations of cycles? First of all, a system should posses at least two degrees of freedom and at least two balance equations of matter (energy, moment of momentum) resources, and descriptions of flows and transformations. For direct and reverse transfers of matter (energy, etc.), being effected in a system, interrelations are necessary and hence non-linear term should be presented in the equations. Linear terms in the equations express restrictions of growth of matter (energy) in internal structures, in other words, a dissipation of a system. In order to prevent synchronisation and loss of degrees of freedom due to interactions, the parameters of equations should satisfy of the self-similarity condition, or characteristic times of system components should satisfy the condition $t_i > t_{i+1}$ **(i= 1,2,...,n-1)**

6.1 Cycles' Functional Models

Since at the present stage of joint investigations of cycles many physical parameters and relations are still conditional and undeveloped, for this purpose it is necessary to be guided by construction and investigation of their functional models, as it was done at the dawn of cybernetics for investigation of biological systems. For this purpose let us consider two functional models that will help to a certain degree reinterpret existing ideas and concepts and to be correctly guided in these problems. The first model represents the simple transformation of matter (energy) given in Chapters 3 and 4. Thus, let us denote by x_i the energy (mass) of the **i**-th component of a system. Then, basing on given considerations, it is possible to write a formal balance equation for system's masses, which has the following from in the simplest case:

$$
\begin{aligned}
x'_1 &= x_1 (a_0 x_0 - a_2 x_2) - (b_{11} + b_{12}) x_1 \\
x'_i &= x_i (a_{i-1} x_{i-1} - a_{i+1} x_{i+1}) - (b_{i1} + b_{i2}) x_i + \nu_i \ , \quad i=1,3, \dots ,n-1 \qquad (1) \\
x'_n &= a_{n-1} x_{n-1} x_n - (b_{n1} + b_{n2}) x_n + \nu_n
\end{aligned}
$$

Here $a_0 x_0 = Q(t)$ represents an amount of initial matter (energy) penetrates into a system within every T unit of due to interactions with a medium. Non-linear transitions in a system are represented by values of energy inflow and outflow at the **i**-th degree of freedom (in the equation). The linear terms represent energy expenditures directed to preservation of the **i**-th degree of freedom ordering $\mathbf{v_i}$ $\mathbf{(t_i)}$ is an energy penetrated into a degree of freedom due to interactions with the corresponding cyclic changes of a medium. Presence of the mentioned three types of terms in the equation should be considered apparently as mandatory. The set equations (1) is specific because: a) each equation has its own characteristic time $\mathbf{t_i}$, and $\mathbf{t_i} > \mathbf{t_{i+1}}$; b) each equation is realised within spatial $\mathbf{l_i}$ ($\mathbf{l_i} > \mathbf{l_{i+1}}$) scale of equilibrium ; and c) appearance of new terms (diffusion, convection, turbulence, advection, whirls and turnovers (see, Chapter 4)) is caused by the rate of preliminary state restoration.

The main question is following: What should be the coefficients $\mathbf{a_i}$, $\mathbf{b_i}$, of Eqs. (1), that energy **Q(t)** transferred to a system within every **T** unit of time, being passed and transformed by its components, be converted and dissipated?

Thus each of the equations of (1) set (degree of freedom) realises a part of the general problem and has its own contribution to a system's stability. **Q(t)** determines a number **n** of equations of (1) set and $\mathbf{a_i}$, $\mathbf{b_i}$ coefficients. There is the following simple relation between **n** and $\mathbf{a_i}$, $\mathbf{b_i}$. If the dissipation coefficient $\mathbf{a_i}$, $\mathbf{b_i}$ in the equations are small, then a number of transformations, i.e. the equations, is big, and if $\mathbf{a_i}$, $\mathbf{b_i}$ are big, then **n** is small. In other words, a length of the transformation's chain depends on **Q(t)** and values of $\mathbf{a_i}$ and $\mathbf{b_i}$. Undoubtedly, determination of optimum values of the monitored parameters has basic significance for the clarification of the system's type and structure. As the investigations show, the total number trophic levels' does not exceed 5 to 6 and a choice of coefficients is determined by rate of a series of biophysical processes and their interrelations. Direction of transformations in a system is provided by coefficient self-similarity, and appearance (vanishing) of a new equation (degree of freedom) is determined by increasing (decreasing) of medium's load **Q(t).** Thus, a main cause of origin of new specification and cyclic changes is an increase of internal load of subordinated system. This principle of origin and preservation of cycle changes in ecological systems is assumed as natural and experimentally checked multiply. It can be of interest only for problems of forming and establishing of the sun system cycles. Application of this principle here means that cyclic changes in the solar system are formed not at ones, as it is assumed by most of researchers, but gradually and stimulating force for this was changes in motion momentum of the central body, in the given case its gradual increase. Optimum condition for origin and evolution regularities of living nature (variation nature) evidences that regularities of structures and interactions origin for nonliving nature and that of its stability preservation should be the and general. Undoubtedly this point of view for cycle origin may help to understand more comprehensively some problems of the solar system (for example, the opposite direction of Venus rotation).

A total number of transformations in geophysical subordinated system apparently are of order of a number of geological systems. Thus the Earth globe, as a subordinated system, and contains from 4 to 5 basic shells, which subsequently transform, convert and dissipate a portion of initial (primary) energy penetrated into them till the next portion would come (Chapter 3). Transformations of the moment of momentum compelled by the centre of galaxy are realised gradually by the primary, secondary, tertiary bodies. Here, the number of transformations is equal to three (Chapter 2). There are five levels with different properties in the world ocean and they, apparently realise the rate of subsequent neutralisation and dissipation of external excitations. Despite the obvious similarity of structures and interactions of cycle changes manifested in various media, many problem concerning them are very far from being satisfactory as of today. Such models, undoubtedly, suffer with one substantial drawback, an excessive generalisation. Proceeding from corresponding physical concepts, it is necessary to define concretely a physical nature of functions and coefficients participating in the equations and their meanings for every system.

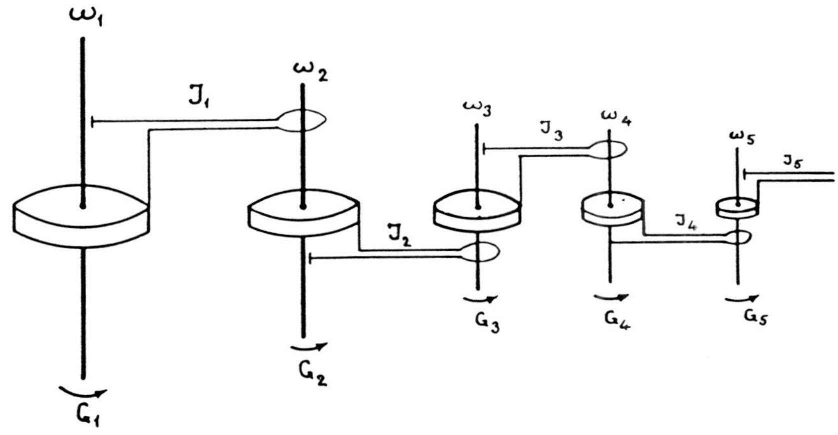

Fig. 19 *Reproduction of the simplest structure and interactions of cycles by means of interacting weak dynamos*

Let us consider another functional model of cycles. It is an electromechanical system and may be of interest in studies of the solar system cyclic changes. Let us discuss a joint of several dynamos coupled to each other by inductive connection (Fig. 19). Each dynamo comprises a magnetic disk subjected to a rotation moment $\mathbf{G_{ik}}$ at every is $\mathbf{T_{ik}}$ (k=1,2,...) instance of time. A disk is characterised by current **I** of a coil connecting dynamos and an angular speed **X** of disk rotation. The dynamic equations of this system represent the laws of energy and angular moment conservation. For the dynamos with **i=1,2,...,n-1** they have them following form:

$$\mathbf{L_iI_i + R_iI_i = X_i (M_{i-1}I_{i-1} + M_{i+1}I_{i+1})},$$
$$\mathbf{G_iX_i = G_{ik} + I_i (M_{i-1}I_{i-1} - M_{i+1}I_{i+1})}, \qquad (2)$$

where the coefficients $\mathbf{L_i,R_i,M_i}$ and $\mathbf{C_i}$ represent inductance of dynamo, circuit resistance, mutual inductance and inertia moment, respectively. The first of the equation energy dissipation inside the system, and the second one describes changes of external effect magnitude. The self-simulation of coefficients means that

$$\mathbf{G_{i+1} /G_i = L_{i+1} /L_i = R_{i+1} /R_i = C_{i+1} /C_i = q<0}. \qquad (3)$$

Let up suggest, for simplicity, that the mutual inductance is the same everywhere, that is $\mathbf{M_i=M_0}$, i=1,2,...,n . The condition (3) is necessary to provide directionality of interactions and extrusion in the system. But it does not exclude a possibility of feedback and influence in the system, which, for example is possible if the internal resistance is negative, the ratio of rotating moment $\mathbf{G_{ik}}$ values within a particular interval of time are not automodel ones. The latter case is important due to the fact that, proceeding from the system stability, deceleration of rotation of one or more dynamos will require, and in the condition of mass deficit, rotation of opposite direction occur.

Using Eq (3) and simple transformations, the set of equation (2) may be reduced to the following form:

$$\mathbf{Y'_i + a_iY_i = b_iX_iY_i \quad , \quad X'_i = G_{ik} - d_iY^2_i}, \qquad (4)$$

where a_i ,b_i and d_i coefficients are rational functions of the parameters of the set of equations (2).

Thus, when the effect of the principal (primary) dynamo on others is consecutive, then their dynamic equations are reduced to the form of Eq (4). It should be only kept in mind that a differentiation for each **i** is carried out with respect to its characteristic time t_i. If there are direct actions of the primary dynamo on the other ones in the system, then the dynamic equations are:

$$L_iY'_i + R_iY_i = X_i (M_{0\,i}Y_{0\,i} + M_{i-1}Y_{i-1}) + X_{i+1}Y_{i+1}M_{i+1}$$
$$G_iX'_i = G_i + Y_i (M_{0\,i}Y_{0\,i} + M_{i-1}Y_{i-1} - Y_{i+1}M_{i+1}) \qquad (5)$$

A simplification of these equation is not clear yet, therefore let us look for periodical solutions of the set (4). For this purpose let us multiply the first equation of the set (4) by **y** and denote, $s= y^2\exp(-2(R/L)t)$. In this case instead of Eqs(4) we get

$$S' = 2ASX \; , \qquad X' = G^1 - BS \exp(-2(R/L)t) \qquad (6)$$

Now, we determine **X** from the first equations and substitute it into the second one, in the result we obtain:

$$S'S - S'^2 = GS - BS^2 \exp(-2(R/L)t) \qquad (7)$$

Let up assume $S = \exp(u+\alpha t)$, where $\alpha = -(2R/L)t$, in this case Eq (7) is:

$$U' = 2AG - 2AB \exp U \qquad (8)$$

According to the conditions of the problem, at any instant of time $t = T_{ik}$ an instantaneous rotating moment affecting the **i**-th dynamo is $G_{ik}(t) = Q_0\delta x(t-T_{ik})$, where $\delta(t - T_{ik})$ is a Dirac's delta function. Such a form of writing means that an external excitation effects only at the cycle origin instant that is the case of resonance is considered. In this case Eq (8) in the new notation will have the following from:

$$U' = -\lambda \exp U + f(t) \; ; \; (f(t) = 2AG_0\delta (T-t_i)) \qquad (9)$$

Multiplying the both sides of Eq (8) and integrating, we obtain

$$U'^2 + 2\lambda \exp U = 2f_0\, U'(T_{ik}) = \lambda_0 \qquad (10).$$

When solving Eq (10) it should be suggested that $\lambda_0 > 0$, and the cases of $\lambda > 0$ and $\lambda < 0$ should be considered separately. Let us suppose that $\lambda > 0$ then in the case $\sqrt{\lambda_0 - 2\gamma\exp U} = \sqrt{\lambda_0}\sqrt{\beta\exp U - 1}$ where $\beta = 2\lambda/\lambda_0$, and the solution is

$$Y(t) = \frac{1}{\sqrt{\beta}\, ch\frac{\sqrt{\lambda_0}}{2}(t + c_1)} \qquad (11)$$

In the case of $\lambda<0$, we obtain

$$(t) = \ln \frac{1}{\beta sh^2 \frac{\sqrt{\lambda_0}}{2}(t + c_1)} \tag{12}$$

which in the initial notation have the following from : $Y(t) = \frac{1}{\sqrt{\beta} sh \frac{\sqrt{\lambda_0}}{2}(t - c_1)}$ (13)

The solution (11) being dependent on imaginary value's $\mathbf{c_1}$, allows on oscillating mode. The solution (13) is decreasing with the increase of **t** , but since both of them depend of $\mathbf{G_{ik}}$ and $\mathbf{I_{i-1}}$, which are changing in time after every $\mathbf{T_{ik}}$ unit, then the solution (11) and (13) depend on the new values of $\mathbf{G_{ik}}$ and $\mathbf{Y_{i-1}}$ and are changing constantly.

In order to determine the integration constant let us set the following conditions. Suppose that $\mathbf{Y(T_{ik}) = Y_0}$, $\mathbf{Y'(T_{ik}) = Y1}$. In this case we obtain from Eq (12)

$$c_1 = \frac{2}{\sqrt{\lambda_0}} arch \frac{1}{\sqrt{\beta y}} - T_{ik} \tag{14}$$

where $\beta = \frac{2\lambda}{\lambda_0}$, $\lambda_0 = f_0 U(0)$, $\lambda = \frac{2AB - 2q(2\beta - \gamma)\gamma B}{LC}$

It may be shown that $\mathbf{c_1}$ is a real number in the case of $\sqrt{\beta Y_0} < 1$. In this case the solution (12) attenuates exponentially. In the other case ($\sqrt{\beta \lambda_0} > 1$) the solution (12) may be a periodical function of the time. Generally, behaviour of **Y(t)** depends essentially on signs and initial conditions. Let us now proceed to the determination of **(t)**. From Eq (4) we have

$$X' = G_{ik}^1(t) - \frac{\gamma q}{c} y^2 = G_{ik}^1(t) - \frac{\beta}{\sqrt{\beta}} \frac{1}{ch^2 \frac{\lambda_0}{2}(t + c_1)}$$

and hence

$$X(t) = \int G_{ik}(t)dt + \frac{B}{\beta} \frac{1}{e^{\frac{\lambda_0}{4}(t + c_1)}} + c_2 \tag{15}$$

If $\mathbf{G_i(t) = G_0 \delta(t - T_i)}$, then determining $\mathbf{c_2}$ in the case of $\mathbf{X(T_i) = 0}$, we obtain :

$$x(t) = \frac{B}{\beta} \left[\frac{1}{\left(\frac{1}{\gamma_0} + \sqrt{\frac{1}{\gamma^2} - 1}\right)^2 \exp(\sqrt{\lambda_0}(t - T_{ik})) + 1} - \frac{\gamma_0}{2\left(1/\gamma + \sqrt{\frac{1}{\gamma_0^2} - 1}\right)} \right]$$

where $\gamma = \sqrt{\beta Y_0}$. From the relation it can be seen that in the cause of $\mathbf{t \to T_{ik}}$ we obtain

$$X(t) = \frac{B}{\beta}\left(1 + \sqrt{1 + \beta y^2}\right) \tag{16}$$

where $\beta = 2\frac{\lambda}{\lambda_0} = \frac{2\lambda}{S_0 \frac{2 y_0}{y_1}} = \frac{ABy_0}{y_1} = \frac{By_0}{2 G^1_{\xi} y_1}$

In the case when $T_{ik} >> \frac{2}{\sqrt{\lambda}}$ (i.e. when $\mathbf{T_{ik}}$ is practically infinite) and assuming that $\mathbf{X(\infty) = 0}$, we obtain :

$$X(t) = \frac{B}{\beta} \frac{1}{\exp(\sqrt{\lambda_0}(t - c_1)) + 1}$$

where $c_1 = \frac{2}{\sqrt{\lambda_0}} \operatorname{arch} \frac{1}{\sqrt{\beta Y_0}} - T_{ik}$

In the case of $\beta Y_0^2 > 1$, c_1 is a complex number. In fact ,

$$c_i = \frac{2}{\sqrt{\lambda_0}} \operatorname{arch} \frac{1}{\sqrt{\beta y_0}} = \frac{2}{\sqrt{\lambda_0}} \ln \rho e^{i\theta} \tag{17}$$

where $\theta = \operatorname{arctg} \frac{\sqrt{1 - \frac{1}{\gamma^2}}}{\frac{1}{\gamma}} = \operatorname{arctg} \sqrt{\gamma^2 - 1}$

and $\rho = \frac{1}{\gamma^2} + 1 - \frac{1}{\gamma^2} = 1$

Thus, we have

$$c = \frac{2}{\sqrt{\lambda_0}} i\theta - T_{ik} = i \frac{2}{\sqrt{\lambda_0}} \operatorname{arctg} \sqrt{\beta y_0^2 - 1} - T_{ik} \tag{18}$$

In this case

$$y(t) = \frac{1}{\sqrt{\beta \rho_1}} \exp i \operatorname{arctg} \left[\operatorname{tg}\theta \bullet \operatorname{th} \frac{\sqrt{\lambda_0}}{2} t \right] \tag{19}$$

where $\rho_1^2 = \operatorname{ch}^2 \frac{\sqrt{\lambda}}{2} t \bullet \cos^2\theta + \operatorname{sh}^2 \frac{\sqrt{\lambda_0}}{2} t \bullet \sin^2\theta$

and again, in the case of $\mathbf{t \to T_{ik}}$ we have an oscillating behaviour with decaying amplitude. From Eq. (19) a period of oscillation, which depends on $\mathbf{t}$ and a solution of a complicated transcendental equation, may be determined.

What can be joined from the consideration of the set (4) in study of cycles? First, the solution of (13) and (19) types reproduce oscillations caused by two independent on each other excitations $\mathbf{G}$ and $\mathbf{Y_{i+1}}$, which in their turn, represent oscillating processes with different characteristic times. Second, similar solutions taken together reproduce basic tendency and properties of natural systems interactions, their drift. An initial portion of matter (energy) passing, following through system's components, is treated and dissipated gradually. Problems of stability of systems with different time (space) characteristic within the frames of the classic theory represents serious difficulties. As in seems, they are essentially connected to distributions of characteristic values and parameters. For this reason, it is required that parameters would be self-simulating and characteristic times would be required in order to obtain co-ordinated oscillations of different frequencies, i.e. solution of (13) and

(19) types, in the system of interacting dynamos (pendulums). In such conditions an uncompensated angular moment is transferred between dynamos within a cycle and provides co-ordination of dynamos interaction and hence a stability of the system.

Then "arbitrary" changes of period and amplitudes in deferent directions are observed in cycle interactions everywhere and always. Periods are permanently in creasing and decreasing at complete "random". The solution (13) and (19) due dependence on $\mathbf{G_{ik}}$ and $\mathbf{Y_{i+1}}$ show such behaviour and are changing "arbitrary" from cycle to cycle. The solutions (13) and (19) represent responds of a dynamo to external $\mathbf{G_{ik}}$ and internal $\mathbf{Y_{i+1}}$ influences. Actually they are dissipate agents for these influences and hence agents of equilibrium restoration. A similarly of the solar system transformations, i.e. gradual transfers of motion momentum, which are spent similarly at every level - to restore disturbed order and expulsion and transformation of momentum - are reproduced principally correctly by $\mathbf{Y_i(t)}$ and $\mathbf{X_i(t)}$ solutions.

But processes restoring equilibrium are not similar always. Thus, the considerations of oceanic cycles show that restoration in different scales may be accompanied by various geophy-sical phenomena such as turbulence, diffusion, convection, advection, whirl or other turnover processes. Just in the same way studies of geological cycles evidence that in tectonic cycles of all orders an active stage is followed by quiet and restoration stage, where different geological phenomena such as slow uplifting of sites, folding, magmatic intrusions, earthquakes and sediment processes make their contribution. Determination of restoration processes for oscillating phenomena is not specifically difficult, it is only necessary to revise an oscillation decrement, in order to find (decaying) oscillations number and attenuation time interval. But they are of principle significance for wave processes, especially for large wave numbers since the restoration processes here lost period (million of years, in the sphere of geological cycle), amounting approximately up to 80 to 90 percent of cycle duration. Hence many geophysical phenomena are virtually and reconstruction processes, and requires individual consideration. In these cases it is more expedient to use the first functional model of cycles by means of introducing corresponding terms into the equations. Despite the mentioned vast properties of the solution (13) and (19), it should not be attributed a deep meaning to them because the set of equations (4) and correspondence of the solar system's parameters and their isomorphism are not revised yet. Besides that Eqs (13) and (19) should be solutions of a variation problem. In other words $\{y_1,y_2,...,y_n\}$ and $\{x,_1x_2,...,x_n\}$ should be defined so that primary (and hence secondary and all other current) motion momentum be distributed between dynamos optimally and be dissipated in the course of cycle . Therefore, the solution (13) and (19) and the corresponding restoration processes should show the certain ability for self-organisation. At lost, it is not clear is the coefficient self-simulation an enough condition for solution self-simulation or not. Nevertheless, despite the mentioned drawbacks, the solutions (13) and (19) play a significant role in studies of cycle structures and interactions.

6.2 THE INNER CYCLES' ANALYSES [1]

The problem of the inner cycle's origin, maintenance and disappearance is important in the analyses of the cycles' interactions. The reasons of their origin, as well as their nature are different. We will attempt to clarify this problem on the base of the special type of cycles (or rhythms).

Cycles' functional models such as (1) and (2) reflect their idealised relations. Moving through the system excitement, in reality, transforms and dissipates incompletely. The energy (substance, exertion, information, signal, etc.) unbalanced allotments are accumulated during the several cycles'. It is converted to a special source of excitement, which is just an inner excitement. In all cases, a balance is established with regard to inner excitement. These phenomena's details still are not so clear. It is difficult to indicate, a priory, the frequencies (or transform stages) where penetrating or transforming energy is used, assimilated and dissipated completely, and in contrary, frequencies - where they take place imperfectly. This question is more prominent and perceptible in the area of biological rhythms. Let us analyse a class of biological rhythms, in the particular, brain γ type of rhythms. Brain rhythms' analyses are convenient for some reason. First, they allow us to understand how and why the high frequency unbalanced rhythms (signals) are accumulated. Seconds, such analyses help develop a deeper understanding and estimate of the nature (focal, diffusion, regional) and origin of the inner excitation (or seizure). Third, they are inspected easily and quickly.

Brain γ rhythms are 4-5 Hz. oscillations of its electrical activity. Some of the individuals brain, or different reasons, around γ type of rhythms spontaneously, and at times create and expedite unusual rhythmic oscillations. They depress and interrupt brain's different sections, zones, domains and centres ordinary co-ordinating functions and interactions (rhythms) and temporarily enforce a new functional process. Being dominant, it is spread by the entire brain. In other words, synchronisation of some rhythms takes s place. Co-ordinated processes such as reciprocal activation of opponent muscle and neurone groups, metabolism (dispatches) and numerous other physiological conversions can be dislocated and replaced by strange interactions and transformations that are controlled from one centre. This produces disturbances, disorder band chaos. Such extraordinary situations are called epileptic seizures or epilepsy by neurophysiologists.

Once again, interest in quantitative methods of research on functional and structural brain problems is increasing. One reason is the vast theoretical potential of strange attractors in non-linear dynamic systems. The theoretical design of strange attractors within the context of neurones, nuclei and other brain structures, and the analysis of their behaviour in phase space has great potential. The significance of non-linear models is especially high with brain diseases when physiological means or remedies are often restricted. Changing the parameters of such attractors within broad limits sometimes allows one to predict and to evaluate physiological and pathological alterations of brain processes. Epileptic seizure is one of the widespread brain diseases, which was analysed in detail by Jasper and others (1969), Okujava (1969). There are epileptic seizure quantitative examinations too; Babloyantz and Destexhe (1986), Kaczmarek and Babloyantz (1977). Epileptogenesis' some discus-sions are expounded briefly by Sarafyan (1993). That discussion is further developed in the present report, which attempts to do some quantitative analysis of data obtained from brain sites. Attention also is paid to brain non-ergodicity, non-linearity, to hidden rhythms in the records, to the separation of dominant channels, and to the origin causes.

Neurobiological background. The data were separated into three different stages; sleep, waking,

1. In this paragraph, we set forth results that we received during the 1992-1993 years at UCLA, Brain Institute.

and disease (temporal lobe seizure). The records were made in the early morning (6:54 a.m.) from a patient and consisted of 2 minutes each of sleep, waking, and seizure stages. At each stage, records were created using 8 channels of EEG with a digitization frequency of 63 Hz. The total data obtained from each channel in each stage was, therefore, 7800 points. A section of the records is shown in Fig. 21. The locations of the recording sites were;

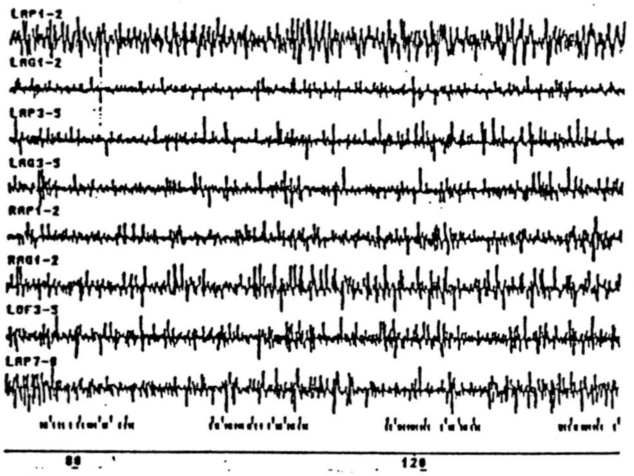

Fig.20 *Normal cortical electroencephalographic activity (EEG)*

LAP1-2, Mesial Part of Anterior Hippocampus (Left Side),
LAG1-2, Mesial Parahippocampal Cortex (Left Side),
LAP3-5, Lateral Part of Left Anterior Hippocampus,
LAG3-5, Lateral Parahippocampal Cortex, (Left Side),
RAP1-2, Mesial Part of Right Anterior Hippocampus,
RAG1-2, Mesial Part of Right Parahippo campal Cortex,
LOF3-5, Left Orbitofriontal Cortex,
LAP7-8, Left Superficial (surface) temporal Cortex.

The following values were selected for investigation: **N = 100, 150, 200, 300, 600,** which represent approximately intervals of **1.5**, 2.2, **4**, **5**, and **9** seconds. The computations are repeated for each **N** until the last interval, that is to say, until the **120** seconds record is finished

Non ergodicity of brain: In dynamic systems there sometimes arise a need to replace "space-average" techniques with "time-average" techniques.For example, when restrictions are placed on access to the spatial coordinates of the systems, then the use of "space-averaging" can became difficult or impossible, and "time-averaging" becomes desirable. Some researchers faced with this circumstance, have attempted to use time averaging in the analysis of brain electrophysiology (Babloyantz and Destexhe1986, Cindy and others 1991, Kaczmarek and Babloyantz 1977). The general reasoning behind their approach is the following.

A dynamic system, like the brain, can be described by a set $\{X_0(t),X_1(t),...,X_{n-1}(t)\}$ of variables which satisfy a system of first-order differential equations. This set might be the values of the EEG at **N** different sites. On the other hand, a differential equation of order **n**, with a single variable $X_0()$ is equivalent to the **n** first order differential equations, where $X_0(t)$ and its **(n-1)** derivatives jointly represent the original set. Sometimes, instead of a set of **n** variables simultaneously recorded from different sites at the same point in time, it is convenient to obtain a set of variables from a single time series recorded from a single site at different points in time $\{X_0(t),X_0(t-\tau), X_0(t-2\tau),...,X_0(t-(n-1)\tau)\}$. The result is thus obtained from a single channel of information distributed over time instead of from **n** channels of spatially distributed information. If the system is ergodic the derived data can be equivalent to the original data. If transformation **T** is ergodic and **f** an integrable function, the "time-average"

$$\lim_{N\to\infty} \frac{1}{N} \sum_{k=1}^{N-1} f(T\ p) \qquad (1)$$

exists and is equal to the "space-average"

$$\frac{1}{\mu(s)} \int_s f(p)\,dp \qquad (2)$$

for almost all **p** (Halmos 1958, Arnold and Avez 1968, Eckman and Rueble 1985). The convergence of a "time-average" to a "space-average" means essentially that the "space-average" can be estimated from a si ngle realization of the process by constructing a "time-average". It also indicates that temporal and spatial distributions of an ergodic process are uniform and tend to have the same structures.This type of approach to structural and functional brain relations is clearly over simplified and unbalanced,inasmuch as, brain spatial and temporal properties are inhomogeneous and not monotonic, even when **n** analysis sites are close.

Non-linearity of brain activity: Brain non-linearity one may verify by a different way. We will confirm brain non-linearity by analysing separate stage records and their interactions. First it was necessary to determine whether the nature of the interactions among hippocampal sites changes as a function of periodic influences such as circadian rhythms. Were the correlation positive, negative, alternating or non-linear and what were their physiological and other interpretations? To justify the independent "time-series" and to measure their similarity, reciprocal double correlation coefficients were obtained from the 8 channels according to the following formula:

$$Rx_ix_j = \frac{N(\sum^N X_iX_j) - (\sum^N X_i)(\sum^N X_j)}{\{(N(\sum X_i^2) - (\sum X_i)^2)(N(\sum X_i^2) - (\sum X_i)^2)\}^{1/2}} \qquad (3)$$

where **i,j = 1,2,3,...,8, i<j.** If many "time-series" are analyzed and each series contains too many terms, the reciprocal double correlation function (3) becomes uncharacteristic for the system. In such cases modified forms (formula 1, in Babloyantz and Destexhe 1986) are used which describe a global or averaged property of the system. However, the above-mentioned type of correlation function is used in the present study for considering the following questions: first, determination of the independence of the channels' information and second, possible alterations of the connections at the passage from sleep stage to disease and waking stages. But apart from the type of correlation function used there is a problem in selecting the appropriate time series length. How long should they be and which duration is optimal for the analysis?. Usually, it depends upon the frequency of the process considered. For high frequency oscillations one should select small values of **N,** and vice versa. In general, it is connected with the classification of different types of physiological rhythms (Pavlidis 1973, Winfree 1980, Sarafyan 1984). These calculations show that correlation functions are changing throughout the sample of sleep. In other words, interactions between hippocampal sites are mostly both excitatory and inhibitory, which should be considered normal and natural. Such tendencies of the interaction among hippocampal sites are also maintained during the waking stage at all **N.** But the pattern is quite different when seizure date are analyzed. In the seizure sample several constant interactions are established among hippocampal sites, depending upon N. Thus, when **N=100** the values **Rx_ix_j, j=1,2,3,..,8, i<j**, are Table 12

1	.000	-.337	.822	.280	-.216	.124	.121	.036
2	.000	.000	-.370	-.396	.508	.032	-.069	-.055
3	.000	.000	.000	.210	-.254	-.083	.189	.092
4	.000	.000	.000	.000	-.657	-.272	.393	.264
5	.000	.000	.000	.000	.000	-.110	-.291	-.045
6	.000	.000	.000	.000	.000	.000	.035	.290
7	.000	.000	.000	.000	.000	.000	.000	-.072
8	.000	.000	.000	.000	.000	.000	.000	.000

These values are maintained throughout the seizure sample. The values of Rx_ix_j at N = **200, 300, 400, 600,** are completely different. Correlation coefficients of the two channel alterations are shown as a function of N in Table 12. The same constant interactions are maintained also between the remaining channels during the intervals considered. What sort of physiologic inference do they allow? This finding suggests relations that conform to the synchronization pattern of the hippocampal sites when their intrinsic frequencies are suppressed and they function under a constrained regime. One can interpret the existence of different interactions among hippocampal sites in the disease stage at the different N's, in various ways. The main cause of different interactions is the influence of the counting length, namely **N**. Indeed, with increasing N, the forms of the correlation function (Fig. 21), show high frequency oscillations which are

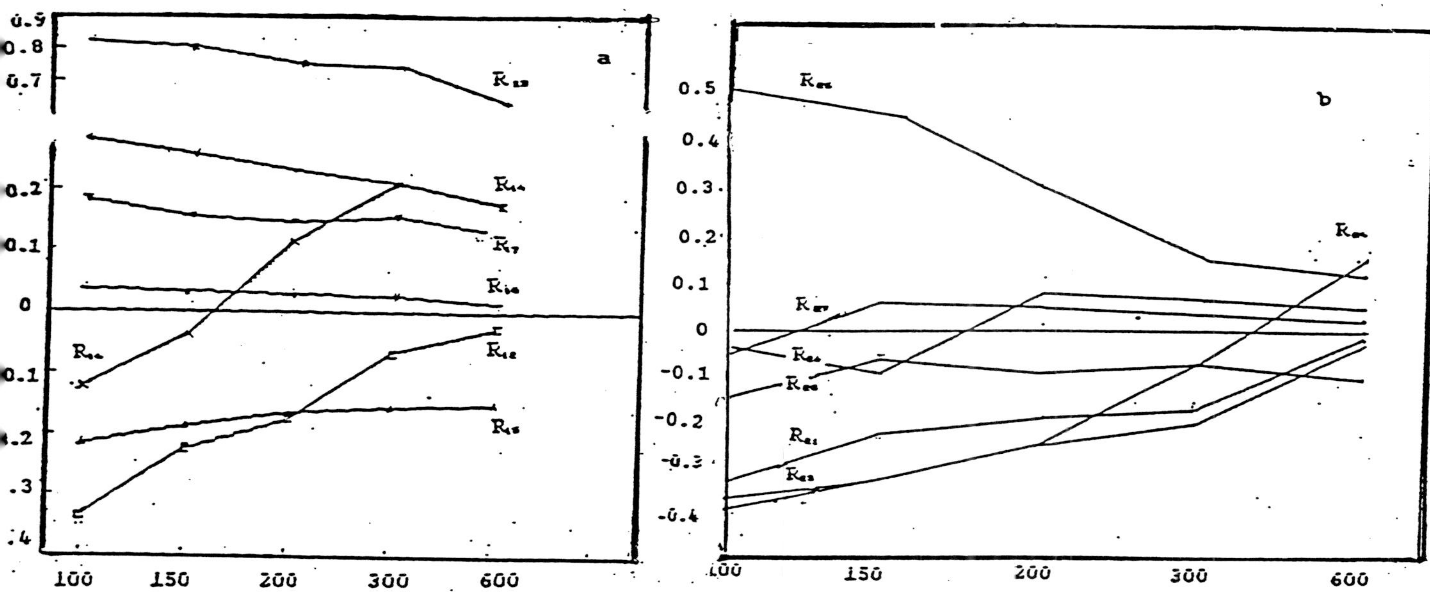

Fig. 21 *Constant interaction at the epileptic seizure dependence upon on N. Interactions Around channel 1 are represented in (a) and interactions around channel 2 in (b)*

gradually damped and low frequency ones which increase at positive correlation and decrease at negative correlation. For example, $\mathbf{R_{13}}$ and $\mathbf{R_{12}}$ are negative and positive correlation coefficients respectively and their values and changes for increasing **N** are as follows

Table 13

N	SEC.	R_{13}	ΔR_{13}	%	R_{12}	ΔR_{12}	%
100	1.5	.822	-	3.9	-.337	-	33
150	2.3	.790	-.032	6.8	-.225	.112	15
200	3.0	.734	-.056	3.5	-.175	.050	31
300	4.6	.705	-.029	4.0	-.068	.107	12
600	9.5	.672	-.033	81.8	-.027	.041	9.0

The values of $\mathbf{R}_{13}$ are obtained by subtracting consecutive values of $\mathbf{R}_{13}$ like this .790 -.822= -0.32 which forms 3.9% of .822 and so on. By the same manner the values of $\Delta\mathbf{R}_{12}$ are obtained from $\mathbf{R}_{12}$.From Table 12. it is easy to see that as **N** increases, positive correlation decrease, while negative correlation increase. Fig.21a and Fig.21b display the data in Table 13. The different behaviour of the curves of Fig.21a and Fig.21b may result from the geometry of the recording contacts or from primary causes, such as the function of the underlying site. In contrast, this tendency of decreasing positive and increasing negative correlation was not observed in the sleep and waking periods. For example, the computations of the first three steps of **600** rows at different values of **N** show that the above mentioned rule occurred only around $\mathbf{R}_{12}$ $\mathbf{R}_{14}$ $\mathbf{R}_{15}$

Table 14

	N	100	150	200	300	600
$\mathbf{R}_{12}$	600	-.046	-.052	-.054	-.065	-.072
	1200	.015	.013	.011	.009	.012
	1800	.250	.225	.027	.020	.016
$\mathbf{R}_{14}$	600	.619	.626	.630	.635	.645
	1200	.612	.614	.613	.613	.612
	1800	.810	.799	.737	.672	.650
$\mathbf{R}_{15}$	600	.120	.104	.097	.091	.083
	1200	.210	.197	.199	.197	.192
	1800	.290	.257	.209	.176	.153

The absence of common regulation of the interactions, in these cases, could be explained by the inhomogeneity of the functions of the different hippocampal sites, and perhaps, by the non-linearity of the interactions. It also indicates important differences across spatial and temporal structures of the "time-series" obtained from the different hippocampal sites. In general, carrying out the computations $\mathbf{R}x_ix_j$ **i,j=1, 2, 3,...,8,i<j,** at different stages and different values of **N** produces small correlation between any two channels, with the exception of channel 3, in which the correlation function is significantly higher than others. Low values of the correlation coefficients indicate interactions between brain sites are almost always non-linear interactions. The nonlinearly of the brain interactions is more clearly observable in the two and three dimension attractors (Fig.24). Indeed, inspections (Fig.24) show attractors (independent of what stage and brain site they correspond to) constantly display cyclic, non-linear behaviours. Low values of the correlation function signify also, that information obtained from the electroencephalogram at the different sites is independent and dissimilar. This property of the correlation functions affirms that the selection of the brain sites was performed correctly. The values of $\mathbf{R}_{13}$ at different **N** and stages are (Table 15)

Table 15.

N	100	150	200	300	600
Seizure	.822	.794	.734	.705	.672
Sleep	.628	.615	.625	.634	.644
Waking	.510	.490	.443	.421	.405

This set of numbers suggests that the records of channel 3 are similar to the records of channel 1. We could apply statistical analysis to the data in order to draw a general conclusion with regard to channel 3. For the present, not every result is explainable. Consider the decreasing $\mathbf{R}_{13}$ on the vertical columns on Table 14, and the positive, negative, and other correlation of the interactions.

Hidden periodicals The calculation of correlation coefficients at different N's, can be extremely useful for discovering hidden periodicals. Actually, from column 5 in Table 12, one can see at **N = 100**, only **3.9%** of interacting frequencies, between channels 1 and 3, form oscillations, whose periods are from 0 to 15 seconds. At **N = 150, 6.8%** form oscillations whose periods are from 1.5 to 2.3 seconds and so on. At **N= 600, 81.8%** of interacting frequencies, form oscillations whose periods are more than 9.5 seconds. So, columns 2 and 5 of table 10 represent overall the periods of the oscillations and their proportions; which form **R** at **N = 100**. By the same manner, one can define the frequencies and their proportion, which form $\mathbf{R}_{13}$ at **N = 150, 200, 300** and **600**. The reasoning remains the same for $\mathbf{R}_{12}$, which increases with **N**. In this case only, the proportions are largest at higher frequencies as shown in column 8. The last could be a reason for a physiological inference. Thus it indicates that around channel 2, cells have higher contribution during epileptic seizure than around channel 3. Otherwise, their function is more "essential" than others. Consequently, we can classify sites, according to their responsibility and specific physiological and functional compositions. On the other hand, a common form of the "time series" **X(t)** is

$$\alpha(t) + \beta(t) + \gamma(t) \qquad (4)$$

restoration where $\alpha(t)$ is a non-periodic function indicating trend, $\beta(t)$ consists of a number of periodic functions indicating oscillation and $\gamma(t)$ is a random variable providing the element of randomness in the series. It would appear that if the problem of isolating and analysing those components of trend, oscillation and randomness can be solved, then the "time-series" will be completely defined and exactly predictable. But that is not the case in the real world. Usually, form (4) represents a reaction of the system, here, it represents a reaction of some brain sites. So the above-mentioned components will vary continuously, with alterations of the external conditions of the system. The problem here is more radical. One should examine characteristic periods for each of the different types of rhythms existing within brain activity, such as the extra-slow, regular alteration of brain electrical potential etc., (Glass and Mackey 1979, Nunez 1981, Winfree 1980). Disregarding principal different scaling time proceedings and interactions led to the superficial interpretation of available facts.

The dominant channel.

The selection of provocative hippocampal sites (channels) which are responsible for the spread of high frequency waves is important. In the present study, optimal search methods were used for the selection of such channels (Bellman and Dreyfuse 1975). The process was as follows: The seizure stage "time-series" was gradually divided and for each interval, $\mathbf{Rx_ix_j}$ $\mathbf{i,j=1, 2, 3,...,8,}$ $\mathbf{i<j}$ was calculated separately. The division process was continued until an interval was reached where the calculated **R** was constant. The channel first exhibiting "invariability" thus was dominant. The above-mentioned method was applied to the 8 channels, obtained at the seizure stage and it was discovered, that channel #5 was dominant: high frequency oscillations initiate first at channel #5. One can also isolate hippocampal sites, which are theoretically responsible for the genesis of an epileptic seizure, via auto-correlation functions and attractors.

Time-dependent properties.

The auto-correlation function was used to discover the internal structure of the "time-series" and their hidden periodicities. The auto-correlation function used the form

$$S(t) = \frac{1}{N\sigma} \sum_{i=1}^{N} X(t_i)\, X(t_i - \tau) \qquad (5)$$

Where, **N** is the length of the considered series, σ is the dispersion, τ is the lag. Interference phenomena between the values of processes **X(t)** and **X(t-τ)** are the physical basis for a very valuable property of the auto-correlation function. It allows one to reveal regular and irregular components of the time series. For **N = 600**, and for **0 < τ< 120** the auto-correlation function was calculated for all 8 EEG channels of the seizure sample.

Results

Calculated auto-correlation functions are basically divided into two groups. In the first, the oscillation is damped due to the periodic components, which can be measured exactly and uniformly. In the second, the oscillation is again damped but this time through irregular components. The "time-series" in Fig.22b are endowed with a wider spectrum than in Fig.22a. Three of 8 of the auto-correlation functions of the "time-series" (channels) are similar and have the form Fig.22a.

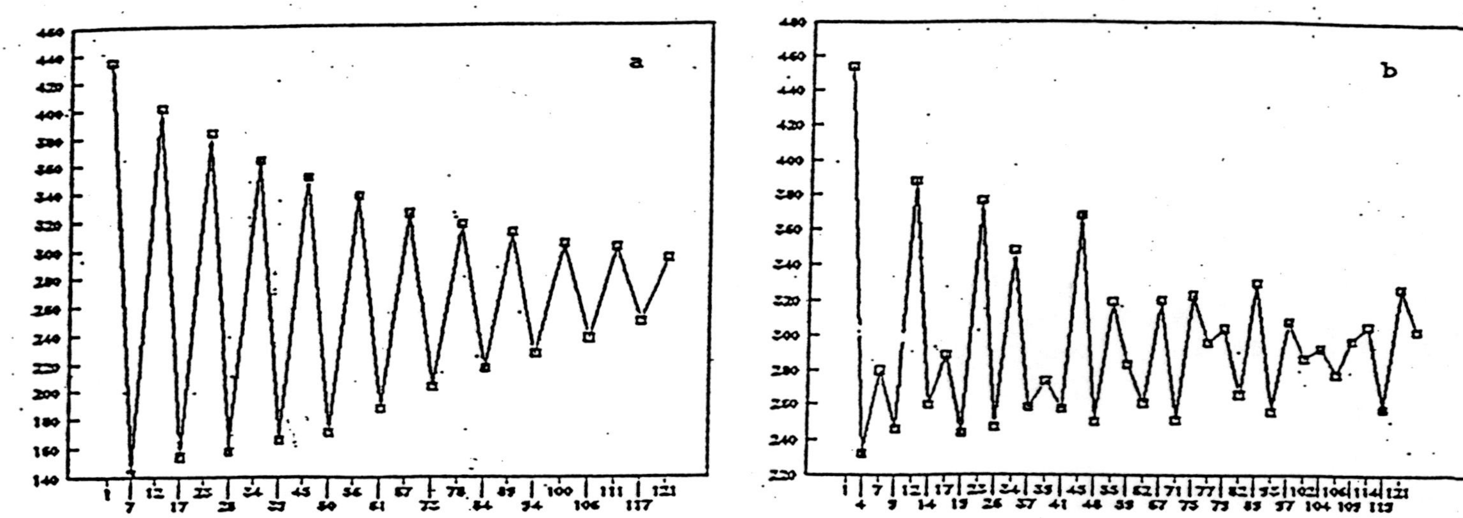

Fig. 22 *Autocorrelation functions S(τ) of the "time-series" dependence upon τ. Interactions around channel 1 are represented in (a) and interactions around channel 2 in (b)*

The other 5 functions have the form represented in Fig.22b. Further inspection of Fig.22 shows that the oscillations in both cases gradually damp around some channels, which in turn, decreases over time. A low frequency rhythm that accompanies the indicated damping, is clearly visible on the autocorrelation functions computed for the seizure stage. Fig. 23 presents one of the auto-correlation functions (channel 2) which include the large period of τ. The kind of physiologic process corresponding to these oscillations has not yet been determined.

One special interest is the examination of two or three-dimensional attractors, which is constructed from the EEG records at different stages, brain sites, values of **N**,τ and other parameters (Fig. 24). Inspection of the attractors shows they clearly are distinguished from each other. They show regular behaviour but distinct configurations, which fluctuated continuously over time. For instance, at the three dimensional analyses, configurations are cross-shaped in the sleep stage, ellipse-shaped in the awake stage, and fish-shaped in the seizure stage. These patterns are replaced by the other patterns as the parameters are changed. Such variations definitely show the inconstant, unsteady relations and inter

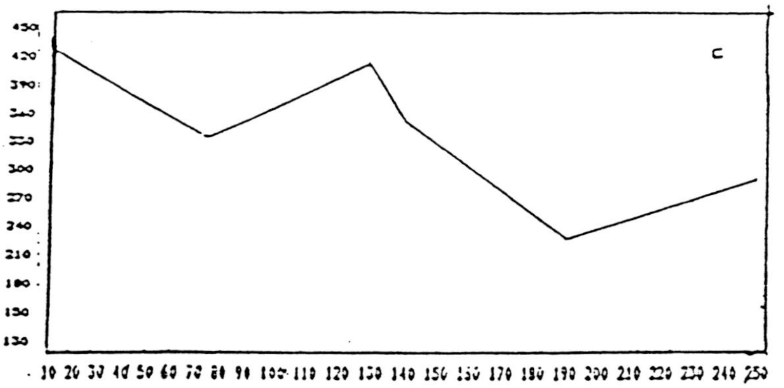

Fig. 23 Autocorrelation function (channel 2) S(τ) when τ = 240

Actions within brain activity, which Would be expected. The alterationa of the patterns indicate also that the we are working with a multiple frequency phenomenon: the basic frequency, at each value of τ, **N** and other parameters, is accompanied by secondary frequencies. Pattern comprehension or their informaivness depends upon parameters' selection order: as far as they are bound or in conformity with physiological (metabolic or molecular) features. The basic parameters, which were used to construct attractors, are:

-H: horizontal viewpoint dimension
-V: vertical viewpoint dimension
-G: path of graphic drivers
-I: input file name
-LR: last row of input, default is 99
-FR: first row of input, default is 0
-LC: last column of a row, default is 0
-1RS: first row shift defines y co-ordinate, default is 1
-2RS: second row shift defines z co-ordinate, default is 2
-CS: column shift defines same data column in row, default is 0
-MXV: maximum value of data in input file, default is 500
-MNV: minimum value of data in input file, default is 0
-XD: x component of eye vector, default is 0
-YD: y component of eye vector, default is 0
-ZD: z component of eye vector, default is 0
-XE: x co-ordinate of eye's position, default is 0
-YE: y co-ordinate of eye's position, default is 0
-ZE: z co-ordinate of eye's position, default is 0

Two-dimensional attractors are determined comparatively by choosing a smaller number of parameters. Several two-dimensional attractors are shown in Fig. 25. Their parameters, for the comparison with the Fig. 24 attractors were the same. All attractors natural sizes in the Fig. 24 and Fig. 25 are reduced.

Discussion

From a physiological point of view there are a number of important factors which currently limit the level of inference these mathematical techniques can provide. Indeed, one can construct two and three-dimensional strange attractors for any channel, for any starting time (point), for any stage and frequency. We could easily investigate the time series' self-similarity, spatial regularity, and other spatial-temporal features. But we cannot make any physiological or metabolic deductions and evaluations from them. The question here is in no means limited to the electrophysiological data, in terms of selecting the hippocampal sites, or in better interpretation of the time series. The question is rather in the extent or manner by which

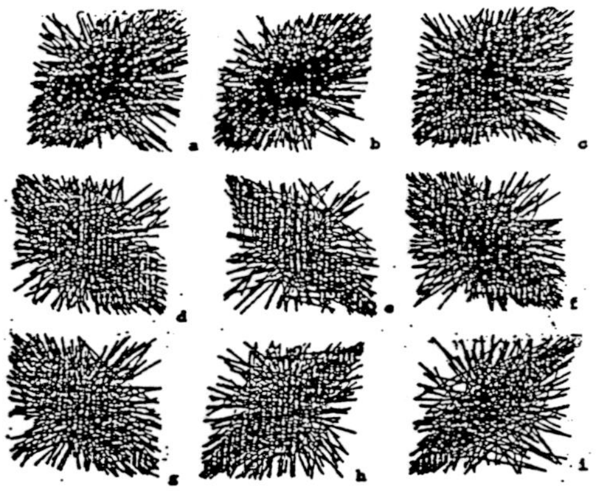

Fig. 24B

Three dimensional attractors of the awake stage.
(attractors are diminished by 90% of their natural sizes)
a.-I0:PRE.TXT -G\BGI\ -CS1 -LC3 -LR1000 -FR0 -1RS12 -2RS12
b.-I0:PRE.TXT -G\BGI\ -CS1 -LC3 -LR2000 -FR0 -1RS12 -2RS12
c.-I0:PRE.TXT -G\BGI\ -CS1 -LC3 -LR1000 -FR0 -1RS14 -2RS14
d.-I0:PRE.TXT -G\BGI\ -CS1 -LC3 -LR1000 -FR0 -1RS28 -2RS28
e.-I0:PRE.TXT -G\BGI\ -CS1 -LC3 -LR1000 -FR0 -1RS40 -2RS40
f.-I0:PRE.TXT -G\BGI\ -CS1 -LC3 -LR1000 -FR0 -1RS50 -2RS50
g.-I0:PRE.TXT -G\BGI\ -CS1 -LC3 -LR1000 -FR0 -1RS40 -2RS96
h.-I0:PRE.TXT -G\BGI\ -CS1 -LC3 -LR1000 -FR0 -1RS72 -2RS72
i.-I0:PRE.TXT -G\BGI\ -CS2 -LC3 -LR1000 -FR0 -1RS78 -2RS78

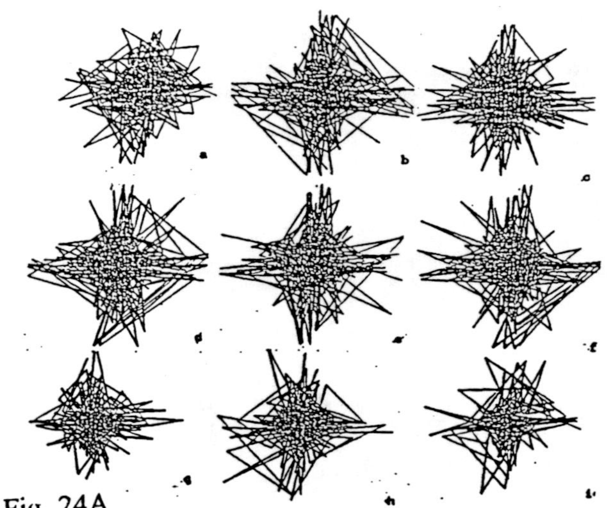

Fig. 24A.

Three dimensional attractors of the sleep stage.
(attractors are diminished by 90% of their natural sizes)
a.-I0:SLEEP.TXT -G\BGI\ -CS3 -LC4 -1RS9 -2RS9 -LR1000 -FR0
b.-I0:SLEEP.TXT -G\BGI\ -CS3 -LC4 -1RS9 -2RS9 -LR2000 -FR1000
c.-I0:SLEEP.TXT -G\BGI\ -CS1 -LC2 -1RS12 -2RS12 -LR1500 -FR0
d.-I0:SLEEP.TXT -G\BGI\ -CS3 -LC4 -LR2000 -FR1000 -1RS15 -2RS15
e.-I0:SLEEP.TXT -G\BGI\ -CS3 -LC4 -LR2000 -FR1000 -1RS20 -2RS20
f.-I0:SLEEP.TXT -G\BGI\ -CS3 -LC4 -LR2000 -FR1000 -1RS25 -2RS25
g.-I0:SLEEP.TXT -G\BGI\ -CS1 -LC2 -LR1000 -FR500 -1RS12 -2RS12
h.-I0:SLEEP.TXT -G\BGI\ -CS1 -LC2 -LR500 -FR0 -1RS25 -2RS25
i.-I0:SLEEP.TXT -G\BGI\ -CS3 -LC4 -LR300 -FR0 -1RS9 -2RS9

Designations in the Fig.4 and Fig.5 are; -LR: last row, -FR: first row, -LC: (or C) last column, -1RS: (or R) first row shift defines y coordinate, -2RS: second row shift defines z coordinate, -CS: column shift defines data column in row, -XD: x component of eye vector, -YD: y component of eye vector, -ZD: z component of eye vector, -XE: x coordinate of eye's position, -YE: y coordinate of eye's position, -ZE: z coordinate of eye's position.

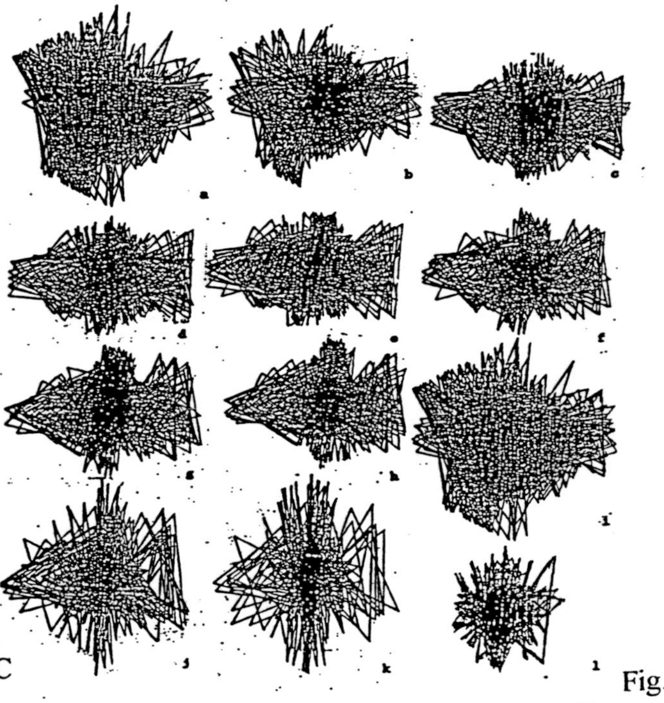

Fig. 24C

Three dimensional attractors of the seizure stage.
(attractors are diminished by 80% of their natural sizes)
a.-I0:SZ2.TXT -G\BGI\ -X0.1 -Y0.1 -Z0.1 -CS2 -LC3 -LR1000 -FR0
b.-I0:SZ2.TXT -G\BGI\ -X0.1 -Y0.1 -Z0.1 -CS2 -LC3 -LR1500 -FR500
c.-I0:SZ2.TXT -G\BGI\ -X0.1 -Y0.1 -Z0.1 -CS2 -LC3 -LR2000 -FR1000
d.-I0:SZ2.TXT -G\BGI\ -X0.1 -Y0.1 -Z0.1 -CS2 -LC3 -LR2500 -FR1500
e.-I0:SZ2.TXT -G\BGI\ -X0.1 -Y0.1 -Z0.1 -CS5 -LC3 -LR3000 -FR2000
f.-I0:SZ2.TXT -G\BGI\ -X0.1 -Y0.1 -Z0.1 -CS2 -LC3 -LR3500 -FR2500
g.-I0:SZ2.TXT -G\BGI\ -X0.1 -Y0.1 -Z0.1 -CS2 -LC3 -LR4000 -FR300
h.-I0:SZ2.TXT -G\BGI\ -X0.1 -Y0.1 -Z0.1 -CS2 -LC3 -LR4500 -FR35
i.-I0:SZ2.TXT -G\BGI\ -X0.1 -Y0.1 -Z0.1 -CS2 -LC3 -LR5000 -FR40
j.-I0:SZ2.TXT -G\BGI\ -X0.1 -Y0.1 -Z0.1 -CS2 -LC3 -LR5500 -FR450
k.-I0:SZ2.TXT -G\BGI\ -X0.1 -Y0.1 -Z0.1 -CS2 -LC3 -LR6000 -FR500
l.-I0:SZ2.TXT -G\BGI\ -X0.1 -Y0.1 -ZC.1 -CS2 -LC3 -LR7000 -FR600

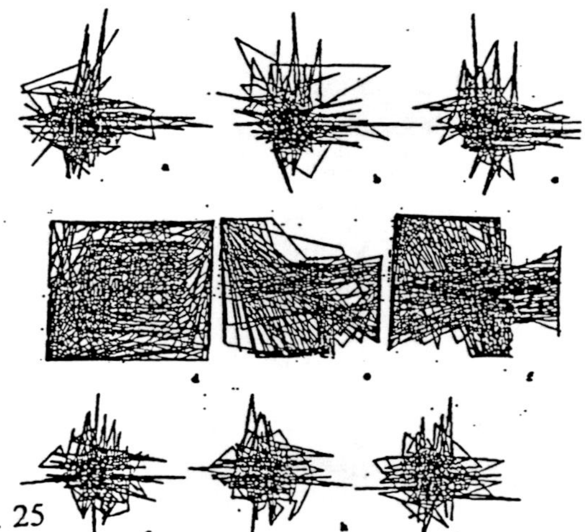

Fig. 25

Designation in the Fig. 24 and Fig. 25 are; -LR: last row, -FR: first row, -LC: last column, IRS: first row shift defines y coordinate, -2RS: second row shift defines z coordinate, -CS; column shift defines data column in row, ID: x component of eye vector, -YD: y component of eye vector, -ZD: z component of eye vector, -XE: x coordinate of eye's position, -YE: y coordinate of eye's position, -ZE:z coordinate of eye's position.

molecular processes generate and initiate a situation, which on the physiological level of the interactions is called an epileptic seizure. When do the molecular (or physiological) processes begin to have an appreciable effect on the physiological (or ontogenetic) processes, or on the other hand, do physiological (ontogenetic) process have any influence on the molecular (physiological) processes? The higher order physiological processes, which are predicted or evaluate by the molecular processes, require one to establish some sort of criterion. Holden, still in the 60's, analysing correlation of the processes by different scales, had noted that "the difficulties of biology result from the necessity to think simultaneously in several scales". In the epilepsy literature the exact and uniform definition of the necessity to think in several scales is still qualitative rather than quantitative. As a result, in present time it is difficult to determine the number of channels which would be optimum for adequate description of a seizure disorder, especially in cases where the origin of the seizure is not obvious.

Despite the above restrictions, such examinations are useful. For instance quick and effective determination, of the efficacy of a medication and its effect on the patient's EEG, can be difficult in a clinical situation. This problem can be resolved comparatively simply and easily by the examination of attractors constructed on sufficient statistical data and information. In the same manner, certain difficulties can arise in the determination of the position of brain recording sites, from which to ascertain the amount of information about brain "disorderliness" such as seizures, etc. Also, this problem can be simply solved on the statistical analysis of correlation functions of the interactions of brain sites (see sec. 2). Generally, the question is not in increasing the resolution or number of "independent" brain recording sites ("independent" by means of recording information). The question is; in the selection of certain brain sites, records from, which will be acceptable for the integrative description of brain seizure features. This problem can potentially be solved by analyses of the following questions: the type of additional information obtained from each channel, the time of appearance of the information and lastly, the possible reiteration of such information. In other words, by the verification of designs based on the adequate statistical attractors.

Finally, analysis of some of the above problems will produce other solutions if the principle of organization of brain rhythms can be definitely elucidated (Winfree 1980, Sarafyan 1986). In fact, what type of subordination exists between brain rhythms? Do low frequency brain rhythms operate under control of high frequency rhythms, or vice versa? The generation of high frequency oscillations or bifurcation among brain pacemakers and establishment between them, suddenly or over time, of a synchronization, should be linked with the imperfect metabolism (or dispatch), with the increasing loading from immediate lower frequency oscillation. This last follows from the organization of circadian and other type of rhythms (Pavlidis 1973, Sarafyan 1986). In particular, epileptic seizures as high frequency oscillations arise as the result of increasing rhythms of the transventricular or intracranial pressure. In order to clarify preliminary causes of epilepsy, and advance the potential of strange attractors for use in electrophysiological research, one should start to examine attractors in low frequency brain rhythms, like transventricular and intracranial pressure rhythms. Such an approach will certainly allow worth-while insights into brain functions and structures. At last, the proposed procedures can be useful for the examination of the self-similarity of any "time-series".

§ 6.3 CYCLES' SELF-ORGANIZATION

As it was mentioned in the preceding chapters, cycles independent of medium their forming being gaseous, liquid or solid are an alteration of processes from equilibrium a state and restoration. The cause of perturbation could be external and internal energy source, with respect to a given scale, and restoration is completely caused by redistribution and reconstruction of inner substance and forces. Restoration and reconstruction processes, their structures are important, because just they, internal regularities, fulfil the tasks arisen for a system; to transform and dissipate corresponding energy (substance) portion due to the some optimum principle. In this sense, internal cycles and their organization is a necessity for stability, with the energy minimum dissipation. Taking account of significance of self-organization in restoration processes (inner cycles), let us consider them more thoroughly.

1. Introduction

Research into the self-organization of objects confirms that properties of self-organization are not only seen in living organisms, but also in many nonliving systems. Self-organization is observed in the earth's atmosphere, hydrosphere and lithosphere. In the above spheres there is a continuous generation of spatial-temporal, hierarchical structures at different scales and as well as many interactions between them. Basically, two types of schemes predominate in the indicated structures. The well-contoured linear and non-linear structural elements of organised, defined scales and regions in the earth's three shells are seen in the accompanying figures. Linear and non-linear structural elements of organization are inherently the same or have similar functional significance, but have different names.

In the atmosphere, the linear elements are called stream currents (Palmen and Newton,1973; Shanina, 1985), in the hydrosphere - frontal currents (Yampolsky,1979; Fedorov, 1979,1983), and in the lithosphere lineaments and break fractures (Shoulst, 1979; Brohakov 1979). Non-linear elements in the atmosphere and hydrosphere are called vortex currents, cyclones and anticyclones (Koshliakov, 1979; Intensive.1984) and in the lithosphere-ringed and spiral waves (Arakelyan,1978; Kratkova,1983). Similar structures and interactions are also observed in galactic formation (Novikov,1981; Pasha, 1981; Peebls,1982; Silk,1983) and in many multicomponent autocatalytic reactions (Zaikin,1970; Win-free,1972; Zabotinsky,1974) etc.

What are linear and non-linear structures, what are their nature what is their functional significance, what is the means by which they are regulated, and what are their generating mechanisms and lifespan within the system? To answer some of these questions, one may consider distinct multicomponent active media. However, other systems, such as the lithosphere, are not understandable and for this, an explanation requires synergetic methods. At the present, the phenomenon of self-organization is seen not only in biology (Ebeling V.1976; Nicolis and Prigogine,1977; Romanowsky, 1982) and chemistry (Farrow,1974; Krinsky 1979; Hess,1979), but also in physics (Oscillatory,1969; Hacken,1980; Nonlinear,1983; Kadanoff,1983; Kerary, 1985; Gorodetsky,1985) mathematics (Turing,1952; Othmer, 1979;Self-organization,1980), geology (Ogorodnikov,1980;Sarafyan,1981:Sholpo,1986) and astrono-my (Eigenson, 1963; Molchanov,1973; Tides,1975; Bazarov,1979; Chechelinitsky 1982). It has been confirmed that self-organization has universal characteristics and is manifested on all of the spatio-temporal scales and interactions material. What is the basis for the different forms of selforganization and what can one discern about the varieties of self-organization when it is considered from a single point of view? In this discussion, some principles of self-organization are explained and an attempt is made to show the unity of their general properties.

2. Definition of Self-Organization

Self-organization is the establishment and improvement of spatial-temporal structures in unbalanced, dissipative media. Self-organization and its development requires only single condition: the existence an extremal, principle which triggers the energy minimum dissipation. From here, it follows

a. The presence of distinct spatial-temporal characteristic scales, or inhomogeneities.

b. The agreement (co-ordination, co-operation) of inhomogeneous interactions.

c. The presence of definite distribution of the heterogeneous characteristics.

The above mentioned three conditions describe the capacity of the system for self-organization. However, one more important condition is necessary for self-organization, and that is the application of energy from an external source. In this regard one may formulate the following statement:

d. The external influence is not only a factor directly generating structures, but also is the creator of conditions for their subsequent generation. This needs some clarification.

The excitation of the medium and its movement takes place neither instantly nor infinitely slowly. Movement of the excitation has finite speed. After a stimulus excites and moves within the medium restoration processes are gradually generated. Restoration processes also have a finite rate, (or time length -τ). In the case of frequent excitation, restoration processes are important and require special attention. In the restoration processes, substances and energy are redistributed between inhomogeneities, which restores the equilibrium of the system. Restoration processes occur when there is a change in the stimulus. The necessity of the restoration processes is obvious; the medium must always be ready to respond, to change, and to distribute subsequent excitements. Usually, different inhomogeneous substances offset each other in the system. When the stimuli disturb the balance between them, then a re-distribution of the substances and of the inner forces takes place. The necessary quantity of energy moves (or is assimilated) to the required site in specific period of time. As a result of similar exchanges and conversions of energy, corresponding reactions of the system are formed. New equilibrium are established and shaped according to the new external conditions. These equilibrium or streamlined exchanges between inhomogeneities are preserved until the next external stimulus infringes upon them. Thus, the rising and establishment of structures depend on the external stimuli and regime that they enter; medium not only acts to suppress an organization and also supports its existence. There are other definitions of the self-organization (Wattington 1947; Ebeling 1976; Nicolis, Prigogin 1977) but one can think the above-mentioned definition is more nearly complete.

3, Relaxation's Features

In active media, specifically, in electrophysiology, the interval of time τ is called a refractory period of the cycle of excitation: in physics-relaxation time: in radioelectronics-a transition period: in hydrodynamics-a turbulent, unstable period, etc.

In mechanics of a solid medium, τ is determined by a relation between two of its characteristics; viscosity and module of shift. In electrophysiological and biochemical processes, τ is determined by the velocity of the inhibitor and activator reactions. In electrical circuits, τ is determined by the product of the resistance and capacitance. The above mentioned determinations of τ testify to the general characteristics of relaxation and to the importance of their research from a holistic point of view (Klafter, 1986). Despite extensive analysis of relaxation in many different branches of science, it is still not completely understood. For example, the composition of relaxation is not entirely understood in economic and social dynamic systems. Even in hydrodynamics, where supposed relaxation processes are clear and comprehensible, the increasing Reynolds number **Re**, does not bring about turbulence if the medium is sufficiently stratified.

At the present time, there are many theories and arguments that attempt to explain or expound upon processes arising within the relaxation period including process of turbulent instability. The turbulent instability problems have been studied in depth, by many leading scientists such as A.M. Obychov, E. Hopf, A.N. Kolmogorov, A.E. Lorenz, L.D. Landau, V. Feigenbaum, etc. Despite different approaches to the solutions of the prevailing problems and the use of mathematical techniques, the results are similar and satisfy all practical requirements. Thus, in the models proposed by Lorenz, sudden conversion from ordered to chaotic currents takes place when there is an increase of **Re;** but in Landau's models, dynamics become complicated gradually. In Hopf's presentations, turbulence rises after four to five a bifurcation in the solutions of differential equations and so on.

A basic idea in the above-mentioned researches is to separate spatial scales that generate turbulence and to reveal forms and mechanisms by which transference and dissipation of energy takes place. These researches, from the formal point view, are correct as far as they go however, they avoid restoration processes and their features. The role of these processes remains unknown for transference and spread of excitation from stimuli that follow each other faster than is the restoration time value of the medium. In the analysis of multiple excitations it is especially necessary to consider the restoration or relaxation properties of the medium. In order to understand excitation processes, and how their characteristics can be revealed, we look, at the spread of repeated the stimulus which are observed in electrophysiology.

4. Relaxation's Significance

According to the existing ideas in electrophysiology (A.A. Vvedensky, M. Braze, A.L. Hodgkin, etc.) spreading excitation in nerve fibbers lasts 0.0694 sec.;but in muscle tissue,0.155 sec. During that time the medium generates a sequence of biochemical reactions which replace one another. Their relative duration is 0.01%, 2.99%, 17%, 30%, and 50% of the excitation period. The indicated phases are: (a) development of the action potential: (b) the absolute refractory period: (c) the relative refractory period: (d) the supernormal or ekzaltacy period:(e) the subnormal phase. The functional meaning and physiological content of each process are essentially different. The first phase defines the time length of the initiating mechanism. All subsequent phases are exchange processes, which are directed to the restoration of the initial energy balance of the tissue. These last phases are labile and sometimes collectively are called the refractory period of excitation. The meaning of the refractory period is as follows; it is a sharp reduction in the excitability of the tissue, allowing the tissue to regain its functions. During that time, the tissue does not respond to external influences that thereby ensures the proper conditions for the occurrence of the internal reaction program until it (reaction) is completed. The specific types of biochemical reactions that are initiated and which replace each other during the refractory period of the tissue are not important for consideration of self-organization. One essential factor is, that after the initiation and spread of each excitation, certain bio-chemical reactions regularly arise and replace each other, and may begin to restore the functional properties of the tissue. Partition of the refractory period into its separate phases and their shifting is not an end in itself but is used to pursue specific goals. The partition of the refractory period solves several principal problems in relation to the object - medium.

First, it protects the system from extra energy expenditures and allows efficient energy utilization, optimizing the actions of the system. Second, the system (tissue) does not have to react to every stimulus. The result is that the system is distinguishable and separates from others. The system becomes autonomous and "independent". The partition into separate phases causes the generation of new inhomogeneous and dissipating structures. The variety of specific functions makes possible the trans mission of definite excitements, and creates different conditions for blocking and limiting "unwanted" signals which separate the constituents of the organization. Third, the system reacts to an external influence not only by setting in motion definitely functional dispatches, but also by producing an obligatory alteration of the functional state. Thus, the partition of the refractory period has principal significance for the generation of inhomogeneous systems.

It seems that the refractory period partition provides all development conditions. But it is not like that. By not taking all consecutive values of excitement into account, the system loses the possibility of the analyses of all chances of development and remains behind the surroundings changes. Otherwise it limits survival chances. Then, what has been the compromise? Certainly, turn time into space, namely go to the structural complexity. Since, at the reduced restoration time the system is able to respond to excitement all values, one should make it shorter. Reducing can be made by various means. First, somehow codify and conserve whole reactions, for use at the further repetitions of excitement. Second, accelerate temporarily biochemical or other kind restoration processes. Third, predict the approaching excitement by internal mechanisms such as rhythms and be active to react to it. With this connection Ugtomsky (1961) notes "A vital process is impossible when some of the partial processes linked with it are not secured." Therefore, like mechanisms are substantial for the surviving process. They are observed everywhere in the bio-systems (Mur Ed, Salzmen 1981) and create some orderliness, which maintenance requires definite energy. Where could that energy come from? It comes from conserved and condensed by the system at the interaction's energy (Odun G. Odum T. 1976). Energy conversion and condensation take place without any premeditation and destination. It is just relaxation extent's result otherwise-system's inner properties that are intended by external conditions again.

Odum E, analysing the organization problems some of bio-systems, concludes that their organization is five times higher than surroundings' organizations (Odum E.1959). This result is interesting for other reason too. It can be a mark for environmental organization and be the extent of the distance between object-media. The above features in the excitability of the active (physiological) media may contribute to a deeper understanding of the equilibrium processes in other systems and may also provide an understanding of details relating to the object-medium since they are not always evident. For example, should arising bifurcation (de-centralization management in economics) always be connected with the upper threshold of excitement (information)?

5. Two Coherent Waves

The excitement spread accompanies by two types of processes in unbalanced, dissipative media. The first process disturbs the existing inhomogeneous connections As a result, regular exchanges and motions are demolished and unstable situations are created which last for a short time. In the second type of process, the unstable situation is subsequently converted to new structural formations and regularities. When the first process takes place there is a destruction and demolition of definite structures, which is accompanied by the liberation and expenditure of energy. In the second process, on the contrary, there is a restoration and revival process that takes place when new inhomogeneous forms are created. As a result, new regular exchanges and motions are established. The generation of these two types of processes can be conveniently presented by two conjugated waves, which follow each other without a gap. A mathematical model describes such waves by two differential equations

$$\frac{\partial \mathbf{U}}{\partial t} = \mathbf{D_u}\,\Delta\,\mathbf{U} + \mathbf{f_1(U,V)}\,;$$

$$\frac{\partial \mathbf{V}}{\partial t} = \mathbf{D_v}\,\Delta\,\mathbf{V} + \mathbf{f_2(U,V)} \qquad (1)$$

where **U** and **V** circumscribe the non-linear connection (for example kinetics) between different physical variables. Thus, in the model of spreading fire, **U(t)** represents the value of energy burned, and **V(t)** represents the necessary quantity of the burning substance. In a model of heart muscle stimulation, **U(t)** is the potential difference across the cell membrane, and **V(t)** is the conductivity of trance-membrane of the cell. In a model of chemical reactions, **U(t)** represents the concentration of the active radicals and **V(t)** represents the concentration of the oxidizing forms of the catalyst. $\mathbf{D_u}$ and $\mathbf{D_v}$ are the coefficients of diffusion; Δ, the Laplace operator; $\mathbf{f_1}$,$\mathbf{f_2}$ represents the interconnection between **U** and **V**. The appearance of $\mathbf{f_1}$ and $\mathbf{f_2}$ depends on the type of concrete system. Each of the equations represents a diffusion type of kinetic equation. If, in the medium (system), the averaging of the concentration proceeds faster in chemical reactions than by diffusion, then one can assume $\mathbf{D_u = D_v = 0}$, and the system equations convert to ordinary differential equations.

Generation of the above-mentioned two types of processes can be displayed also by convection, advection turbulent, current, vortex motions, and by other energy-exchanging processes. The rising of such processes at the interactions depends on the external and internal conditions, and on the parameters of the system. For instance, restoration and balance by diffusion requires a long time for large size systems. Consequently, faster restoration processes are necessary. Diffusion again will be converted, say to convection if the intensity of the energy becomes critical for the system. Restoration and balance by the convection (more so, by diffusion) likewise does not affect media, where there are periodical accumulations of energy (tension, strain) in the separate sites. In such cases more rapid energy-exchanging processes are necessary, like-vortex currents. Vortex currents (cyclones and anti-cyclones) quickly move accumulated tensions long distance from their rising sites and establish balance, which widely are observed in the atmosphere and hydrosphere (Lorenz 1967; Fedorov 1983). Finally, restoration processes can take place by turbulent currents or movements when the system relaxation time is longer than the time between impulses. (Hydrodynamic Instabilities ... 1981)

So, relaxation, dimension, distant between excitements and other parameters stipulate motion type. Their alteration makes it necessary to adjust to different functions or their variations. The most diverse function must be adjusted. Consequently, the appearance of convection advection, vortex currents, turbulence and other energy exchanging forms do not change the basic concept: interactions describe by two conjugated waves the first of them represents excitement spreading processes and the second one-restoration processes. Above-mentioned all phenomena as well as blood, bile, hormonal, humoral, chemical, electrical effects, and others are only coupling means by which the heterogeneous are communicated, integrated, and therefore, realize the balance.

6 . Research Two Directions

What can one say about the behaviour or in other words about the work of the medium, which is described by the system differential equation (1)? Is it possible for repeated and incessant passages to exist between wave processes? If so, then what kinds of conditions are necessary to explain concentric, spiral or other types of self-preserving patterns and structures? The answers to these questions basically depend on two factors; the characteristic of inhomogeneity, and the nature (regime) of the incoming excitations.

Let's consider a simple example. Suppose the medium is homogenous and its relaxation is τ. Assume, also, that the spread and motion of the excitement in the given medium takes **T** units of time, but the length between any two excitations is constant and equal to **r**. The variety of the arising dissipative structures depends on the relation between parameters τ, **T** and **r**. Thus if $\mathbf{r = T}$, then, after every **T**, one can observe in the medium repeated free passage of excitations and wave motion processes. When $\mathbf{r = T}$, rising patterns and structures will be identical and will repeat without interruption; the medium works by its maximum rhythm, as a new excitation comes just at the end of the restoration process of the previous excitation. All sites (points) of medium constantly pass from relaxation to the excitation stage, avoiding the resting stage.

Similarities between the patterns and structures are disturbed whenever **T** becomes greater or smaller than **r**. In the case **T** < **r**, patterns are again freely repeated but they are already fulfilled every **r** unit and by delay **r-T**. Among the parameters τ, **T** and **r**, the most interesting case is **T>r**, or τ >**r**, namely when the length between excitations is smaller than the period of relaxation. In such cases, the relaxation processes prevent the free passage of subsequent excitations: patterns (movements, exchanges) become disorderly, irregular, and chaotic. The situation intensifies when the medium consists of two or more heterogeneous, and **r** becomes variable.

When the subject under investigation is the objective law of the spreading solitary excitement (wave) then the experts' attention is concentrated only on the equation, such as the first one of the system (6.3.1). When the peculiarities of the spreading waves are established and their distortion, dispersion and absorption are defined, one can evaluate the extent medium inhomogeneity. In a similar approach system activity, its inhomogeneity and structures usually are secondary. What then are only considered is the order of the spreading waves and, perhaps, the development and evolution peculiarities of the waves in different boundary and notable initial value conditions. These questions are examined by theory waves. When the subject under investigation becomes the system, its structural and functional steadiness, position, manner and mechanism of dissipation, alterations, then one should consider equations such as the second one of the system (1). The stimulus is considered given, and one is seeking mechanisms, which provide the conservation or stability of spatial-temporal relations of the system. Theory of stability and control just analyse steadiness, conformity conditions of system spatial-temporal connections.

7. Orderliness and Disorderliness

As was already noted, a complex analysis of the phenomena arises when τ > r. In similar cases it is very difficult to separate object-medium system's components. And apparently when combining solutions of two equations of the system (1) one can not say something concrete about the characteristic behaviour and evolution of the systems. Hence, the presence of arising chaos is of tantamount importance in the research of self-organization. From the formal point of view, chaos arises at the alteration of system parameters, supposing that external influences are fixed. However, does the alteration of parameters actually reflect mechanisms, which are generating chaos in the natural systems, or is there something artificial? The answer remains unclear. Furthermore, by which inhomogeneous are regulated? How is the geometry lifetime or chaotic nature organized?. One may attempt to answer some of these questions using the above-mentioned description.

Consider again the spreading processes in an active medium. According to the above notion, chaos arises by the execution of the condition where τ>**r**. In biological systems the refractory period τ is labile and is connected to and regulated by external stimuli. When the condition τ >**r** arises, at once the refractory period begins to undergo change: the relaxation processes are accelerated and, as a result, τ becomes shorter. Otherwise, the condition τ >**r** is converted to the τ <r: bio-system moves away from a stressed state, renews its normal functional dispatches (in chaotic situations, the system expends more energy and spends more time to carry out its function than in a normal state). Therefore, in living systems chaotic behaviours are not preserved long; due to the parameters' liability, regular functional dispatches are substituted to the chaos, stress. Thus, rising chaos is limited by the discrepancy between the alteration rates of system internal and external parameters. In computer simulations disparities are created by the changing dynamic equation's coefficients (namely by altering internal parameters) until they become criiti-

Fig. 26 *Multispiral reverberators in the chemical active media (after Zhabotinski)*

Fig. 27 *Large scale's non-linear structures in the atmosphere*

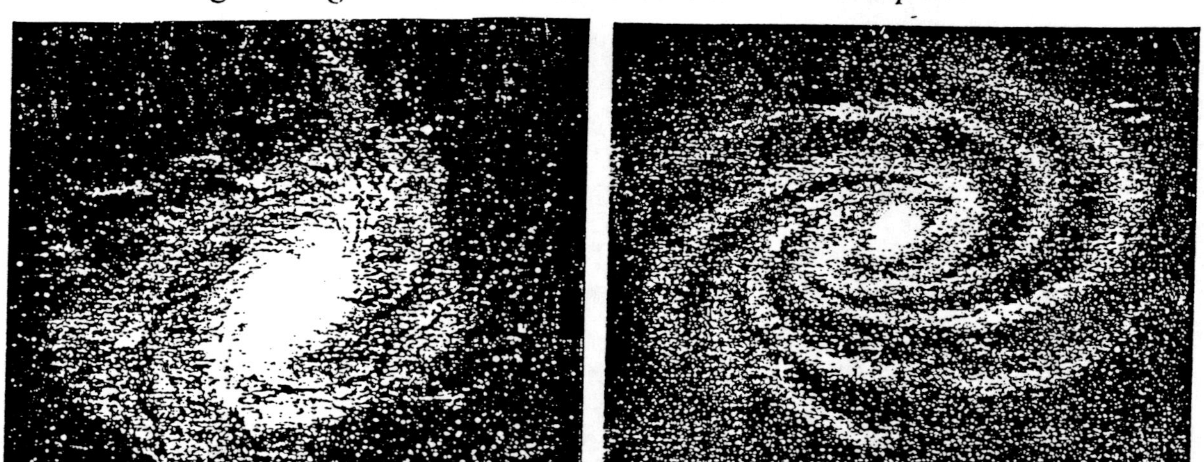

Fig. 28 *Non-linear structures in the space*

cal. The physical explanation of the critical values of coefficients is, seemingly: system is unable to conduct and dissipate the given excitation. In a biological system discrepancies arise as a result of increasing the load of external strain. Initially, bio-system characteristics (such as excitability, refractivity plasticity, etc.) are changed slowly with respect to the increasing load of the medium, allowing the system time to avoid a chaotic stage and time to spread excitations. However, when the external load becomes protracted and durable then changing limits (moderating temporarily restoration processes speeds) becomes insufficient and chaos generation already is inevitable: the necessity of the conduction of functional dispatches and the provision ensuring equilibrium of the exchange processes generate new differentiation (physiological labor's division). Within the system, new functional units (inhomogeneous system) are formed, which are able to effectively transfer and dissipate in creasing load in time. As a result, there is a doubling (or a joining) of some of frequencies and rhythms (Sarafyan ,1984).

Finally in the generation of turbulent movements in hydrodynamics one initially assumes that the only condition is adhesion, namely the treatment to zero of transitional motion on the border (boundary). Then one more condition is added, the roughness of streamlined bodies Probably, in both cases the above analyzed $\tau > \mathbf{r}$ condition should be added. Means of the condition $\tau > \mathbf{r}$ is clear and is not connected to any special spatial scales: disorders can arise within every scale of the interaction, if excitements in the considering scale follow each other sooner than is its relaxation time. The condition $\tau > \mathbf{r}$, certainly, can explain the existing the large-scale turbulent movements in the atmosphere, in the hydrosphere and in other spheres.

8. Structures' Self-Hierarchy

The excitement (tension, rotation moment) is not distributed uniformly in the inhomogeneous solid medium. It is accumulated and placed by certain directions, forming cellular structure. Tension higher values becomes cells' edge (borders) intersections. Accumulated in the intersections tensions later are spread by spiral shape, establishing balance in the medium. As research shows cells' inner chemicals (mechanical, physical) properties are distinguished from each other.

These results are well known for the atmosphere since 60's years (Lorenz,1967; Palmen and Newton, 1969), for the hydrosphere-70's years (Maksimov,1970;Fedorov,1979) in the lithosphere such reasoning became more common since 80's years (Shoullts,1979; Sholpo,1986). Contoured in the lithosphere spiral and ringed formations distribution, like cyclones and anticyclones distribution in the atmosphere and hydrosphere, is not random (Arakelyan,1978; Kratkova,1983; Earth's,1987). They are trajectories by which tension, accumulated at the intersections of the same scale lineaments (linear forms)is spread (Melekestev,1979; Sarafyan and Arakelyan 1981).In the lithosphere restoration processes also are multistage, polyphase. That components are rising, folding, volcanic (rheologic), tectonic processes, earthquake, sedimentation and other geological processes, which replace each other during the geological cycles. The lengths are changed in time too. In the atmosphere and hydrosphere relaxation compositions (different scale motions) still are not elucidated. Actually, the same phenomena are scan in the cosmos. Cosmos also has cellular, hierarchical structure; it is quantized and there are many spiral-shape, ellipse-shape galaxies, having their source from intersection thread-shaped (linear) structures (Silk,1983; Peebls,1982; Novikov,1981; Pasha, 1981; Toomre,1977; Eigenson,1963). Existing in the solar system functional (structural) asymmetry is an additional form of galactic labor (loading) partition.

On the other hand, in active media such as the heart, excitement is spread by spiral-shape trajectories (Winner and Rosenblueth,1946; Balahovsky,1965; Burkov,1974). The same trajectories meet in the crystals' dislocations and other solid bodies (Reed 1954; Zamorsky,1965; Rheology 1962; Kerary, 1985) etc. So, non-linear formations shown in Figs 26, Fig. 27, Fig. 28, and Fig. 29, are trajectories spreading excitations in the inhomogeneous media. In the atmosphere the lifetimes of these formations' last 3 months, in the hydrosphere the lifetime is 3 years (Koshliakov, 1979), in the lithosphere it's several orders higher. Lifetime spiral galaxies are measured in billions of years (Ambartsumyan, 1964; Toomre, 1977). In contrast to the above in the autocatalytic lifetime generating structures are much shorter, for that reason, they are called dissipative (Zaikin, 1970; Krinsky, 1972; Farrow, 1974; Zabotinsky, 1974). So, we see the space everywhere has similar structure. Basically, its functions (reactions) in the corresponding medium are similar too. From the evidence above it follows that everything in nature is subjugated to the same organization principle, which should be expected.

Inhomogeneous generation in biology, such as cell, muscle, tissue, organ, organism, specie, association etc. is an inevitability for the changeable external conditions (loading) and is directed to the energy effective conversion or utilisation. But, inhomogeneities' origin is not endless. It is possible, up to now coupling means are able to unite all inhomogeneous (frequencies) as a whole, with a common goal and to establish balance in time: if labour's division is constantly integrated. According to Shmalhouzen "Differentiation and specialization inevitably get in contact with the subordination and integration." So, in order to form a whole, inhomogeneities' spatial-temporal parameters should be subjugated or selected by specific order:

Definition. *Embedded inhomogeneities, such as the above, interacting in the space, we should say, have self hierarchic structure if their characteristics are obtained from one other by the self-similarity transformation.*

Medium and system structures are linked by the system functions. As much separate function is partitioned with regard to medium factors, as interactions with the medium proceeds capaciously. Contrary to this, reacting to the medium definite factors, the system must have conformable structure, which just realizes interaction with the medium element. If nature found some successful solution to biological problems (and seemingly, not only of biological problems) then it preserves that forever through evolution. Proceeding from this supposition, one could think, self-hierarchy of the structures is applicable to any developing and evolutionary system. For instance, consecutive formation of the earth's shells (Sarafyan 1996) and particularly, primary, secondary, tertiary and other order bodies in celestial mechanics and astrodynamics as result of change to external, galactic loads. Existing relations between the above bodies somehow confirm such relations. Thus, the velocities of rotation of several planets are similar and this is suitably represented by the relation $\mathbf{A=k\ M^{2/3}}$, where **A** angular momentum, to the unit mass of a planet, is connected with of its rotation plus angular momentum its satellites, **M** mass, **k** constant (Tides...1975). At the same time, the velocity of exchange, calculated to the unit of an organism's weight linked by its size: $\mathbf{Q=aW^k}$, where **Q** quantity absorption of oxygen during the unit of time, **W** weight of an animal, **a** constant, **k** power (Schmidt-Nelson 1984). Usually, **k=3\4.**

9. Organization's Collapsing

As was already noted, organization's main purpose is to establish balance with regard to external excitement, timely converting its influence (energy). The conversion of the external excitement's energy is conditional. Since, dependence of the organism's functions on the conditions of the external medium is determined mostly not by the stimulus' physical strength, but the "biological" significance, that the stimulus acquires for the species in the phylogenetic and individual development process. So, external excitement's nature can be various depended on the level of organization. They are variables or **cycles**, among which organization establishes a balance. Wattington (1947) defines organization levels by the sudden generation

Fig. 29 Non-linear structures in the Earth's crust (after Sholpo)

of quiet new properties of the elements included in the composition of the organized whole.

The limits of the changes of the organization structure, in which its functional correlation with the surrounding diverse mode excitement is not infringed, experts call it "adaptation norm". The adaptation norm embraces all species and organization levels, irrespective of that, which ecological niche they occupy. Like restrictions exist in physics and other places too, where they have diverse names. If external influence, load excels to the adaptation norm, then organization is collapsed. There is another, more radical cause of the organization collapsing, but sometimes that is avoided reasonably. Thus it is known that the organization's properties slowly are changed at the interactions. These invisible changes gradually prevent and deflect the conversion of external influences, which decelerate the course of regular functional dispatches of the organization. If deflection or deviation is not renewed (like "tissue" excitability is retrieved permanently by the restoration processes behind each excitement) then in the result of incomplete, conversion processes, the organization, after a certain time will collapse. In that result, all work will be unwarranted. A question arises: how to protect the forming orderliness, which realizes the effective conversion of energy? How to counterbalance the inevitable deflections of organization in order to preserve the starting purpose? If one puts a lot of work into the organization, disregarding its price, then one can generate some mechanism that compensates functional deviations of organization, and in time, pre-determinedly, realize the above-mentioned conversion of energy, and therefore, conserves the acquired orderliness. As such mechanism one can consider self-reproducing processes

In the end let's consider a simple optimization process that illustrates the origin of some of the above-mentioned issues and relations Let's consider the maximization operator

$$\max(X_1, X_2, X_3, \ldots X_{n-1}, X_n) = X^* \qquad (2)$$

Let's assume, that this problem arises at the moment of time **T, 2T, 3T...** Consecutive, pair-wise comparison of (2)'s elements is a standard method of finding the maximum of **n** constants. If each comparison needs one unit of time then the total time for the determination of X^* will be **(n-1)** units. If **n-1>T,** then a question arises. How does one to organize the finding process in order to diminish this time length? The obvious way is to partition original set into **m** subsets and apply the same procedure simultaneously to the subsets. Simple calculation shows that the computation time for the determination of X^* in this case will be

$$n/m + m - 2 \qquad (3)$$

units. A question arises; for which **m**, is (3) minimum? The extremal points of (3) are $m=\pm\sqrt{n}$, and it represents a parabola, with the concave up. So, at the value of $m=\sqrt{n}$, the minimum of the function (3) is $2(\sqrt{n}-1)$.It is easy to see that $(n-1)>2(\sqrt{n}-1)$.

If $2(\sqrt{n}-1)>T$ then the same question arises again; in which organization does the time length $2(\sqrt{n}-1)$ diminish? Let's divide the starting **n** constants set into **n/2** parts, and define simultaneously the greatest element in each of them. Obtained **n/2** constants set let's again divide to the biggest extent namely to the **n/4**, and define simultaneously the greatest element in each of them, and so on, until elements number in the last subsets becomes 2. Simple calculation shows that in this case total time for the determination of X^* is **k**, where **k** is determined from the $2^{k-1}<n<2^k$. Obviously $k<2(\sqrt{n}-1)$. The letter manner of maximum finding is notable, since the corresponding time extent can't be diminished more; it (as well as, division too) is final and limit. In other words the last process finds the maximum in the smallest time. So depending on the values of **T, n** the structures gradually are changed. Does the battle for survival act by the same way?

The more complicate case is, when **T** is different for each of $X_i, i=1,2\ 3,..,n$. In this case the pair-wise comparison of (6.3.2)'s elements, obviously is meaningless. Interesting questions arise if the functions $\{X_i(t)\}$ are not independent and T_1, T_2,T_3,...are not accidental. Particularly, the sets $\{X_i(t)\}$,in order to be united and serve a common gaol is necessary 1.the periods of $X_i(t)\ i=1,2,3,..n$ are satisfied to the conditions 1.$T_1 > T_2 > T_3 > ... > T_n > 0$ and 2.the functions $\{X_i(t)\}$ weakly are linked (Sarafyan 1993).

10 . Conclusion

When properties mentioned in the section two are created and established, then medium demonstrates a competence of self-organization; in the medium spatial-temporal structures are constantly generated and improved depending on changes in energy sources. Inhomogeneous arises through the doubling or joining some of the frequencies and rhythms, due to energy, substance optimal distribution or utilization. After doubling or joining, organization begins to interact in addition with the excitements in respect to it was inert, passive before. Therefore inhomogeneous is improved permanently through the replenishing new rhythms and corresponding structures' orderliness.

Condensed energy is used under unfavourable conditions: protecting prior rhythms, creations and orderliness. Actually by the energy redistribution during the time its better more effective utilization takes place. Later just due to the prior rhythms and structures, the system's passive actions gradually are converted to the active. Its sphere of influence increases depending up on circulation of energy (substance, information). Indeed, in the former times' rhythms reduce restoration time, and reactions began formed faster and sooner than before. As a result, the system becomes able to interact to many more energy sources (diverse mode excitements) surrounding and increases its chance of survival. The only restriction is; there is not more requirement with respect to energy source.

In contrast, in the conditions of unprotecting former rhythms and orderliness, balance establishing becomes mechanical; it will be repeated at the first over and over again namely all necessary reactions to the excitement will be formed, builded all over identically, without any promotion. Time extent, spent for the restoration and chance of survival will be the same forever. The rhythms' formation that reflects the external likewise alterations is the most prospective mechanism of the reducing of restoration times. In the starting phase, rhythms are originated by means of substance or energy sources alterations. Later due to energy or substance optimal using, instead of its sources' variations, information about them is counted and considered. Thereto, rhythms are formed already by the information or signal (like conditional and unconditional reflexes in the electrophysiology). So, rhythms' generation directed to quickly establishing balance could have an energetic, information base and manifests certain hierarchic (subordinated) structure. Moreover, as more single rhythm is subdivided with regard to the environment certain factor, as the balance with that factor is integrity: the organization utilises more distinctive, individual and particular potency of the external conditions.

Summarising, one can say, the minimization of the restoration time in the changeable external conditions, with energy's effective utilization stipulate of the self-organization origin, which level changes over time. Self-organization level is characterized not by energy dissipation or conversion extent, but by the rhythms, by their composition, which have been formed up to that time. As rhythms interaction sphere, otherwise freedom's order, is wider as the organization level is higher: since it utilises more "energy" (diverse mode excitements) and therefore, the establishing balance is more capaciously. The functions' and structures' self-hierarchy prevails in a distributing and concentrating systems. The property of self-hierarchy is prevalent also in nature and in complicated media. However, many problems of self-organization are still not clear nor understood (for example, development's exact motivating forces, bifurcation reversibility conditions, the satisfactory conditions of self-organization origin) and require further analyses. The discovery of new general properties of self-organization allows a deeper understanding of the conditions arising from self organization and of the mechanism that regulate it.

6.4. The Cyclic Organization and Astronomical Some Issues

1. Two Important Inferences

Joint analysis of cycles testifies that the tension in the inhomogeneous media, such as the atmosphere, hydrosphere, and lithosphere is not distributed uniformly. Strengths are accumulated and moved in certain directions, forming a cellular fabric. The movements of tension is not continual in the fabric, it is cyclic. The preservation of the fabric depends on forces that feed. Further, the cells' inner properties are distinguished. The tension's extreme values become intersection points to the cells' borders, clinging onto the circulating tension's portions to come across one another and collide, or they move away from one another. In both cases the instability originates in the medium; in a case the huge accumulation of matter and energy and in the other -its colossal deficiency emerges. In order to preserve balance, amassed energy within a node should be scattered or spread somehow. In the non-homogeneous media, the excitement spread and movement usually takes place in a spiral shape. The spiral-shape movement is also formed as a result of the rising around the nodes' angular momentum. What triggers the angular momentum, is the energy portions collision (or remoteness) around the nodes'. Thus, the function and role of the linear (fronts and currents) and non-linear structures (cyclones and anticyclones) are well known in the inhomogeneous media. The cyclones simply remove and transport amassed energy from their nucleus to outskirts, establishing temporary equilibrium. In the case of an anticyclone, the energy and matter movement and transportation have opposite directions. They are collected and transported from the outlying vicinity to the node (which is the vertex of the spiral), where a tension's magnitude is the lowest. Again temporary equilibrium is established as a result of the transfer of mass-energy to the depths, to the inner levels of a hydrosphere or atmosphere. Generally, a spiral-shape movements a substances of different levels are constantly mixed and an equilibrium quickly (in comparison with other forms of mass-energy transfer: advection, convection, turbulence, diffusion, thermal conductivity and so on) is founded in the medium[1]. The cyclones are the only mechanism that quickly transfer mass-energy and connect the medium's farthest regions together. Whereby, temporary equilibrium is established in the spaces of large scale. Hence it follows that the time length of establishing the equilibrium is very important. To be changed the reactions of the system must simply be formed faster then the external force is able, to changed. Both types of spirals can permanently be formed and developed, while the concentrated energy doesn't become scattered (or contrarily, accumulated), while the averaging of the matter and tension density doesn't become established in the medium. Since spirals realize energy conversion, transmission, and dissipation within given intervals (cycles), they are extreme trajectories. It means, the medium (system, organization) is subjugated to the principle of least actions-PLA: the energy's changes in time and space, are optimal.

The second inference from the cycles' joint analysis is: a cycles' structural and functional organization testifies that them spatial and temporal distribution is not random. Each cycle describes an

1. The radiation also is a coupling means for the balance. Its origin is usually connected the transportation of energy in the shortest time.

irreversible process. Linked irreversible processes form a common direction, which indicates the route of energy transformation and transportation. Connected cycles produce self-organization and development. The cycle–l_1 with length t_1 under PLA can generate the stable cycle-l_2 with length t_2, if it is protracted and $t_1 > t_2$. It means, the cycle l_2 never generates a stable cycle l_1 with $t_1 > t_2$, even if the l_2 is more powerful then the cycle l_1.Further, a system with the cycle l_1,(interacting or balancing the external force f_1 only) after initiating a new cyclic alteration l_2, such as $t_2 < t_1$, also starts to offset commensurably with the f_1 external force f_2, with regard to this force the system was indifferent in the past. Connected cycles form a spatial and temporal hierarchic fabric, which have components embedded, particularly to satisfy the condition

$$\mu_n >> \mu_{n-1} >> \mu_{n-2} >> \ldots >> \mu_2 >> \mu_1 >> \mu_0 \qquad (1)$$

Where, the constant value $t_i = 1/\mu_i$, intrinsic in the system's **i-th** level, is much smaller then is the constant value $t_0 = 1/\mu_0$ which is typical for external forces. This presents the apparent outcome: the reactions of the system (or its components) are formed and accomplished faster than the external excitement (force) can be changed. Otherwise, the speed of **i**-th level relaxation processes is more than the speed of **(i-1)** - th level relaxation processes. Therefore, each level is distinguished by its spatial and temporal structure, precisely by its energy conversion, transmission, and dissipation executed by the cycles. In this regard Ugtomski noted rightly in 1880, that evolving systems are leaned toward the subordination of rapidly reacting components with slowly reacting components (Ugtomski 1961). We have meet that for first time in 1968 when the multilevel stochastic autonomous' behaviours were analyzed (Sarafyan, 1968). The importance of (1) condition has been note by Haken in 1978. In dealing with the sub-process of **i-th** level, the **(i+1)** and **(i-1)** levels' influences one usually is considered as large and, as small parameter respectively. The entire study of the (1) is insufficient yet. Thus, the stratification and hierarchy of the space depends on energy.

2. How does Cycles' Organization Correlate with the Solar System?

The above-mentioned phenomena have their analogies and parallels with some astrophysical issues. Seemingly, the analogies are not only manifestation of physical processes but also laws of mathematical form and expression. Thus, similar phenomena are contoured in the cosmos: the universe also has a cellular, hierarchic structure. It is entirely quantized. Circulation of the fabric matter and the energy, depends on certain conditions, coming across one another and encountering in the intersection points, or contrarily – move away from one another. In the first case, matter's huge conglomeration emerges, in the second - substance's colossal deficiency originates. Proceeding on the nature of a nodes' different sign or direction tremendous angular momentum arises starting at a critical value of amass energy. It permanently generates and develops a spiral-shape movement around the nodes, which transports amass energy from its centre to remote, outlying areas. So, numerous spiral-shape and ellipse-shaped galaxies emanate around the intersections of threadlike structures. The ordinary spiral scatters and spreads out mass energy from its nucleus far away to the outlying spaces. Contrarily, the barred spiral captures and gathers huge interstellar dust, clouds to the centre of the spiral, where a void prevails, where there is an enormous deficiency of matter and strength. In both cases different levels' of massive interstellar material is mixed and commingled through the spiral-shape mass-energy transferring in the shortest time. As a result a temporary balance is established comparatively in large scale's spaces. As a balance mechanism, the spiral (ordinary and barred) galaxies are extreme trajectories, and so, they are inevitable structures in the universe. The spiral structure disappears when congregated energy is scattered and transported far away from its origin and density's and matter's averaging is established within the given interval (cycle). So, energy transformations and interactions in the universe are also subjugated to the principle of least actions, which was expected

From above-mentioned reasoning one can justify:

1. why and how the rotation of planets take place?
2. what triggers a mass and size of the planet.
3. to predict evolution's general direction of changes of orbital parameters.
4. the depth of hierarchy (primary, secondary, tertiary).
5. the conversion of energy to mass beyond the equation $E= mc^2$.
6. the parallelism between self-sustaining, self-propagating, and self-organization mechanisms.
7. the function and role of visible and invisible mass in the life of galaxy.
8. what triggers mass cyclic transfer in binary systems.
9. the inverse rotation of Venus, as an collateral form of balance.
10. what triggers weakly cycles (comets).
11. the gradual origin of stellar systems, such as the solar system.

By some sense, the analyses like of stereotypes seam meaningless, but it is not so despairing. The new conception is creative sufficiently and long-term from the cycles' organisation point of view.

3. Motions' Distribution in the Solar System

On the base of our conceptions of cycles, let's briefly substantiate the last statement: the solar system's gradual origin viewpoint. We ourselves should begin especially with the last question, hoping that in the course of seeking an answer we shall throw light upon the other issues. Astronomical at least two objective laws are favourable for that viewpoint. The first law relates to the spatial-temporal distribution of the revolutionary and rotary angular momentum of the primary, secondary, and tertiary bodies. The second reality is the objects' revolution and rotation directions and orientation.

Let's note by $\mathbf{E_0}$, $\mathbf{E_i}$, $\mathbf{E_{ji}}$ and $\mathbf{R_0}$, $\mathbf{R_i}$, $\mathbf{R_{ji}}$ revolutionary and rotary angular momentum of the primary, secondary, and tertiary bodies respectively. Proceeding from the definition of the angular momentum one may obtain known relations (Alfven, Arrhennius 1976, MacDonald, 1975)

$$\mathbf{E_0} > \mathbf{R_0},\quad \mathbf{E_i} > \mathbf{R_i},\quad \mathbf{E_{ji}} > \mathbf{R_{ji}} \quad \text{and} \quad \mathbf{R_0} > \sum \mathbf{E_i},\quad \mathbf{R_i} > \sum \mathbf{E_{ji}} \qquad (2)$$

These relations testify that the primary, secondary, and tertiary bodies' angular momentum yield by order to each other, and that there is an alignment and coincidence between them. The angular momentum's magnitudes are cyclic: during the time they reach their maximum and minimum values that are changed from cycle to cycle. Further, the directions of the primary, secondary, and tertiary bodies revolutionary and rotary momentum is identical. They have the same, common orientation, which testifies to the presence of a co-operative end.

The similarities of spatial and temporal organization among revolutionary and rotary motions of the primary, secondary, and tertiary bodies testify that these alone form a system with a united purpose. In other words, the revolutions' and rotations' current distribution in the solar system is the result of definite conditions (forces). The same relates also to the average movements' distribution. In this regard Goldrache notes "The prevalence of the quasi-measurement (synchronic) average movements in the solar system is inconsistent with the supposition of random distribution planets' and satellites' average movements" (Goldrache, 1975). The solar system is an organized dynamic system, with the components having definite functions. The meaning of these functions will be clarified below. The only condition that consolidates all movement and mass in the solar system is the angular momentum of the galaxy centre and its equilibrium in certain periods. Precisely, the share of angular momentum, of the solar system, consecutive and continuously is transfer and transition from sun to planets, and from planets to satellites during the galactic cycle. Namely, there is a non standard cascade transformation. This inference follows from the analogous processes existing around the cycles of other scale.

4. Equilibrium in Multilevel Systems

As has already been noted in Chapters 3 and 4, in the atmosphere, hydrosphere, and lithosphere the equilibrium is realized through the cascade transformations: large scale motions are continually converted into small scale motions until they reach the limit of turbulence, where the motions are converted to heat. Otherwise, small scale' motions conduce and serve to stabilize large motions. That is why some high frequency oscillations permanently accompany low frequency oscillation; they simply consume and expand the excitement's energy, and therefore create conditions for the normal flow of low frequency oscillation. High frequency oscillations together represent exclusively relaxation processes for low frequency oscillation. So, the equilibrium of low frequency oscillation is impossible without them. The low and high frequency oscillations distinction is that at any inconsiderable change of the low frequency oscillation, the high frequency oscillations have time to deviate from their state of equilibrium and restore many times. So, in the non-homogeneous systems the equilibrium represents a multistage process: it is continually accomplished by the energy's consecutive transitions from low frequency to the high frequency oscillations. It is furnished by forming a hierarchic fabric, which has a lifetime that is much longer than the excitement's transition time. This concept is characteristic not only for the hydrodynamic systems but also to any other multilevel system.

Likewise, the functioning spatial-temporal relationships among revolutionary and rotary motions of the primary, secondary, and tertiary bodies evidently serve the same, common purpose. That is to say, the angular momentum equilibrium is fulfilled by cascade transformations within the solar system. In this respect Alfven and Arrhenius note " There is not any doubt, the momentum pump occurs between planets and satellites" (Alfven, Arrhenius 1976). The pump of angular momentum exists also between the sun and planets, between the galaxy's centre and sun. The only difference is the nature of the transition of scales. In the hydrosphere they are comparatively unstable and unpredictable, whereas in the astrodynamic systems they are stable and definite. The number of the transition scales herein also does not exceed 3-4. The small motions (short length cycles) serve the large scales motions (long length cycles), by constantly balancing the portions of energy (angular momentum) transported to them and so on. How is this original fabric constructed, preserved, that it realises angular momentum cascade transformation?

5. Angular Momentum and Accretions

The capture, accretion, aggregation, and contraction of matter in space and the time do not occur randomly. Usually, their origin is connected with gravity. Gravitational waves are ripples in the overall geometry of space and time that are produced by moving masses. This argumentation is vulnerable. Observation of the scattering, dispersal of interstellar dust and clouds show that gravity is not pervasive and accretions do not occur permanently, throughout space. So, there is a specific trigger for accretion. The accretion of interstellar substance occurs by two causes, first in the increasing of the galaxy centre's angular momentum, and second, as a result of maximisation of total angular momentum (more effective quantization of mass). The accreted material equilibrates the changes of the primary (secondary, tertiary, or other order) angular momentum. The accretion simply forms new structure, which function is to neutralise, liquidate arising additional angular momentum. As a result, a new equilibrium is established. So, new accretions are assigned permanently for overcoming the restriction appended to the magnitude of the transfer of angular momentum during a given cycle. The structures emergence or origin is due to the exchange of the energy and mass with the surrounding media.

Similarly, if for some reason the galaxy's centre compelled angular momentum has had an additional increase, then the capture and accretion of substance, of surrounded the solar system, take place. In contrast, if a compelled moment of momentum decreases in the galaxy's centre then some substances of solar system gradually move away; it is assembled around a more powerful and substantial moment of momentum. The forming of the tertiary (secondary or primary) body depends on the optimal ways of the balancing moment of momentum. Which way is the compensation and equilibrium of the additional angular momentum optimal within given intervals and constraints, such as the angular momentum on the unit of mass of planet, connected with its rotation, plus angular momentum of its satellite: $\mathbf{A= kM^{2/3}}$ [6]. Thus, to have a secondary (a primary) body with a certain angular momentum (mass and motions), or to have a small secondary and the tertiary bodies with the same total angular momentum, which quantization of the angular momentum is optimum, in the given temporal and spatial restrictions? This problem by its character is similar to the maximum problem discussed in **§3.** How do we partition the **m**, in order to achieve a maximum of **A** in the given time interval? Particularly, a given angular momentum **A =1,2,3,..,J, J-1,..,2, 1** transmit during, **T =1,2,3,..,N** in galactic cycles, when the unit of accreted mass and its circular velocity respectively are **m =1,2,3,..,M**, **v = 1,2,3,..,V**. Which configuration of the balance is optimal?. In order to gain time, in **§6.3** one was proved that initial consecutive multistage a process were replaced by parallel multistage processes. When the structural and functional diversity (configuration) allows us to obtain the problem exact solution. How is it manifested in this case? Let us consider this question circumstantially.

6. Restrictions Overcoming Mechanisms

Every time the restrictions arise, setting limits to further cycles' changes, such as centre of mass and evolutionary (rotary), the problem counterbalancing of angular momentum is overcome by introducing a diverse sort of novelty. For instance, changes of the structure (substance's regular accretion) and changes of the corresponding cycles (centre of mass, revolutionary and rotary). This concept is universal: it is applicable in any media and spatial and temporal scale. The reverse rotation of planets such as Venus also was directed to the balance of the changes in angular momentum's changes. There are the mass-energy' various opposite circulation in the atmosphere and hydrosphere too, and experts are convinced, that such stable circulation have a definitely share in the system's equilibrium. They are the ordinary consumers of energy, particularly angular momentum, as is the "normal" circulation. The inverse rotation mathematically means instead of addition there is subtraction. Thence, it follows that the primary angular momentum's portion intended for Venus, was decreased. This can be the result of the angular momentum's redistribution within the solar system; the already formed Venus, in order to conserve its stability, reacts and equilibrates the new ("decreasing") angular momentum by the inverse rotation. This explanation is controversial since such cycles are still singular. In the inverse circulation (rotations, movements) one can connect also with the mass' deficiency, in regard to the total angular momentum intended for the system. It is correct in small viscosity conditions though. Known hydrodynamics experiments confirm that [25, 27].

7. Angular Momentum Quantization

From a given angular momentum's optimal distribution we can say a satellites' revolutionary and rotary momentum is the effect of the restriction of the planets' revolutionary angular momentum. Satellites' revolutionary and rotary motions compensate and balance the angular momentum compelled by the corresponding planet. Otherwise, there is moment's permanent transition from planets to satellites. As, investigations show "The tides are the basic mechanism that pumps angular momentum from the planets' rotation to satellites revolutionary momentum" (Goldrache, 1975b). Alfven and Arrhenious assert the same too. The satellites' equilibrium represents a third-stage balance of forces. It means that they were formed after planets, due to of quantization local angular momentum.

The planets revolutionary and rotary angular momentum is the effect of the restriction of sun angular momentum. According Goldrache sun tides are too small for the angular momentum's pump from its rotation to planets' revolutionary momentum (Goldrache, 1975a). Here there is the other mechanism for the momentum's pumps. Usually, the revolutionary movements are defined by the orbital parameters. The changes in a planets' energy (angular momentum) are exposed in the variations of orbital parameters such as semi-major axes-**a**, eccentricity-**e**, orbit's inclination to ecliptic-**i**, and so on (Goldrache, 1975). The eccentricity makes a complete pulsation between $\mathbf{0} \leq \mathbf{e} \leq \mathbf{0.053}$ during the **92000** ± **3000** years, for the inclination, and the precession they are $\mathbf{42^0} \leq \mathbf{n} \leq \mathbf{48^0}$, during the cycle **21000±8000,** and $\mathbf{21^0,08'} \leq \varepsilon \leq \mathbf{24^0,04'}$ during the cycle **40000±2000** respectively. The sun's rotary angular momentum's influence on **a, e, i** and on the other orbital parameters, as well as interaction between them, is confused and understandable although incomplete. Despite the sun's rotary angular momentum is one order more than the planets' total orbital angular momentum, nevertheless it can not be trigger for the planets orbital parameter changes. Their responsibilities are basically the sun's revolutionary angular momentum and its fluctuations as long term cycle. Taking into account cycles' organization, one can assert that planets' motions compensate and equilibrate the angular momentum transferred from the sun. The sun rotary momentum's influence on the planets orbital parameters is sub-ordinate. The planets' equilibrium represents a second-stage balance of forces. It means that they were formed after the sun, due to quatization of regional (with regard to the solar system) angular momentum.

The sun's revolutionary and rotary angular momentum, as well as centre of mass cycles, are effects of the galaxy's centre angular momentum. Its revolutionary and rotary motions after their formation compensate and equilibrate continually the angular momentum, compelled by the centre of galaxy. The sun equilibrium represents a first-stage balance of forces within solar system: it means that the sun was formed first of all in the solar system, due to quatization of the global (with regard to galaxy) angular momentum. Evidently, the rank of organized celestial bodies, should start with asteroids and comets, as weaker equilibrium mechanisms. Thus, there is optimal quantization of the angular momentum and its further cascade transformation, dissipation. The angular momentum exchanges are investigated more weakly than the other of types of exchange processes. Matter circulates unlike energy: the energy transformations go by in one direction, distinct from the substance's cyclic motions.

The cycles of tides pump and transfer the planets' rotary momentum to satellites. The cycles of the planets orbital parameters pump and transfer the sun's angular momentum. The galactic cycles and its deviations around their mean value (**200 million** ± **20 million** years) at the geological axes of time, pump and transfer the galaxy centre angular momentum. As noted earlier, the cycles are extreme trajectories (system reactions) which are changed depending on external forces. So, the solar system in like manner is an adaptable dynamic system with an external force. This is the galaxy centre's angular momentum. Every step of the angular momentum change is a direct result of the previous step. Hence, each irreversible change in the solar system's composition is a direct result of the previous step. This is the analogy of the evolutionary principle of continuity. The angular momentum's optimal quantization and further its permanent transfer generate within the solar system (and not only of solar system) a spatial and temporal hierarchy: organization in time is just as a important as organization in space. So, the hierarchy's emergence does not end in itself. It is a necessary and satisfactory condition for establishing equilibrium in the shortest time (see §3).

8. The Forming and Collapsing of Spiral Structures

The above reasoning directly follows from the features of angular momentum development. Indeed, the coming of tensions in the fabric nodes do not take place at once, instantly. They occur gradually, with definite speed. The start of tension accumulation, a (critical) value of an angular momentum, arises around the nodes. Concurrently, the structures begin to be formed and congregated, while starting to realize an energy conversion, transportation, and dissipation. As noted in §3, between the function and structure of the dynamic system, there is an intimate relationship and interdependence. They are changed and evolve simultaneously depending upon external conditions and the principle of least actions. In further aggregation of tension, the angular momentum progressively exerts and extends, becoming a more and more powerful source of energy. The structures are improved, transformed, and inflated conformably. As it is noted in (Binney, Tremaine, 1987) " The interstellar gas seems to be an essential ingredient of spiral structure". They are differentiated and bifurcated becoming more and more vast, ending with their own functions (moment of momentum its share). This is an evolutionary process; a new functions and structures are permanently added to the old functions and structures, by mean of accretion and bifurcation. So, when angular momentum is changed, then previous structures and configurations become unstable. This instability leads, due to the principle of least actions, to the formation new spatial-temporal structures, which shield, and change, modifying former structures. As a result the recent external changes (additional angular momentum) are balanced. The system constantly reflects external forces' in all fluctuations in order to be stable. Such mechanism of the equilibrium is not newness, it is well known in dynamic systems. Further, angular momentum reaches its maximum extent, and is preserved for a short time. The dynamic systems, such as the solar system, reflect optimum to any external changes. As already noted, when restrictions arise, setting limit to further change in cycles, like the centre of mass or evolutionary (rotary), and the problem of global (regional, local and so on) angular momentum compensation is solved by the substance primary, (secondary, tertiary and so on) accretion. So, within the galaxy, gradual quatization of the angular momentum is fulfilled; from global to regional, regional to territorial, territorial to local and so on. Namely, the galactic labor's division (the angular momentum's a bifurcation) continually takes place. As a result, the stability of a stellar disk is acquired by the rotating differentially. Generally, balance equations must be investigated, particularly "spiral structures through the interaction between the orbits and the gravitational forces of the stars of the disk, and thus should be investigated using stellar dynamics"(Binney, Tremaine, 1987). So, all necessary functions (for example, the revolutionary and rotary momentum of entangled bodies of the solar system) are fulfilled with maximal conformity to the current angular momentum during each galactic cycle. That is to say, angular momentum is transferred continuously, and total transfer of the angular momentum is equal to zero during the complete galactic cycle (Goldrache,1975b). The solar system evolves and operates in an energy and time deficiency and in limited loading conditions We can say that the solar system evolves to the stage of maximum compact packaging with regard to the given angular momentum. Therefore, the solar system was directed to a definite end. The well-organized mass' partition and collective motions serve it exclusively. Over and above the solar system has developed in itself from start to present as a system. Thereafter, the opposite direction's process takes place: angular momentum gradually decreases. The structures decompose, fade, and lose their significance. Spiral galaxies totally vanish after the exhaustion of the strength's central source. In the atmosphere it is well known that the lifetime of the spiral structures or spiral-shape movement lasts three months. In the hydrosphere they last up to two years. In the lithosphere the spiral formations function several million years. Therefore, a spiral galaxy's lifetime also has a finite quantity, and takes place in certain compulsory stages: origin, infancy, adulthood, old age, and death. Hence, in space the solar system's external forces are not a given amount, but are variables, and are subjugated to the motion equations of long living system's. One can say, long-length cycles are spatial structures for short-length cycles, on the contrary, short-length cycles are temporal structures for long-length cycles. In this connection, Haken points out that long-living systems subordinate themselves to short-living systems (Haken,1978). Whatever preserves the life of the galaxy as a system is more essential than what preserves the interaction of the solar system. The transgression of the angular momentum transition between the centre of the galaxy and solar system leads to the collapse of the galaxy much faster than does the transition of angular momentum between planets and satellites.

9. Visible and Invisible Substance Function in the Galaxy (Radiation as a Balance Mechanism)

The restrictions in connection with the revolution's slow character can be overcome by the rotary speeds sharp increase to the origin of radiation. So, the restriction with the slow speed angular momentum transferring is overcome by radiation. This means, when quantization of **m**, and cycles (revolutionary, rotary, center of mass) don't have time to balance the angular momentum, then equilibrium's the fastest mechanism, such as radiation, emerges and is formed. The radiation transfers amassed momentum all over, and transforms it in the shortest time. The emergence of radiation noting changes the system's goal. In the galaxy (or its part) seen and unseen substances together continue their functions: converse, transmit, and dissipate the angular momentum their shares or portions. The same phenomena we see in the lithosphere: energy conversion, transmission, and dissipation in the tectonic active zones (geosinclinales) are several orders higher than tectonic passive zones (platforms) (Khain, 1973). It does not prevent the lithosphere to be a system, which interacts permanently with the mantle and rotary momentum, and offsets corresponding excitements. In the atmosphere and hydrosphere there are also numerous zones of the same signification. The more noticeable example of diverse conversion, transmission, and dissipation of external energy in a complex system of different parts are the living organisms. Thus, distinctions in the metabolism in different organs are tremendous. In more active organs such as the brain, heart, and kidney the metabolism's intensity is almost a hundred times higher then inactive organs such as skin. It does not prevent them to be assembled and surviving as a system in the changeable medium. Like relation-ships basically exist between invisible and visible components of the galaxy. Between them there is certainly an allocation and transmission of angular momentum. Thus, the primary cause of rising radiation, is the necessity of the fastest way to transfer angular momentum. The transfer of momentum carried by radiation to an object on which the radiation falls we observe everywhere and continuously (Binney, Tremaine, 1987). This is related to entire ranges of spectrum. By this connection it is worth indicate a huge, important, and relatively undeveloped area of investigation when different type of radiation (such as γ-rays, **x**-rays, visible rays, microwave, radio and so on) are converted continually to each other. So, the radiation constitutes equilibrium at the shortest time, but the thermonuclear reactions just are by-product.

10. Galaxy Angular Momentum and its Gravitation

The only force that controls spatial-temporal dependency within the nonlinear structures (the cyclones and anticyclones) in the atmosphere and the hydrosphere is angular (revolving) momentum. What forces prevail beyond of the nonlinear structures do not matter. The important thing is, that inside of a cyclone the angular momentum is the determinant, and it dominates any other force. The same determinative rule is within the spiral formations in the lithosphere. Apparently, the galaxy's own angular momentum, and distribution also determine the spatial-temporal relations among structural elements of the spiral galaxy. If even there are some forces within the galaxies, distinguishing from angular momentum, then evidently they are all produced and subordinated to the angular momentum. Within a spiral galaxy, the angular momentum prevails and permeates everywhere, whatever the equilibrium mechanism. Moreover, spiral galaxies are formed just under an angular momentum. As the angular momentum predominates everywhere within the galaxy and is manifested in the all scales and balance mechanisms, it performs the same role that the gravity does. Actually, they both display the same attributes, they are both omnipresent, all over, leaving the same effect on everything within the galaxy. So, one may assert, that the gravitation within the galaxy is its own angular momentum. This inference is essential since it enables way one to know more about the genesis, composing and the limits of the gravitational waves. Now, one can say why should attraction or repulsion exist? Attraction simply increases total angular momentum. Below we will show mathematically why it is needed.

11. Drift of Spiral Structures

The facts testify further, that the spiral galaxy's kernel moves away from its cradle, and can collide or cross over, other galaxy, depending on whether or not their kernels are united. It is unknown what forces control these events. The movement of the spiral- structures as a whole is not unique. Analogous phenomena-advection are observed permanently in the atmosphere, hydrosphere and, also in the other medium. Matter and energy amassment, as well as their spreading, and scattering are cyclic around the fabric nodes. Similarly, the void and emptiness emerges around the nodes' and its replenishment by matter and energy is also cyclic. Each node can be in one of these states: positive, (when it scatters mass-energy: the node is the nuclei of an ordinary galaxy), negative (when it accumulates mass-energy: the node is the vertex of a barred galaxy) and neutral (when none of them occurs). Consequently, during definite time interval (long length cycles) the some nodes of the fabric are the nuclei of ordinary galaxy, some nodes are the vertex of barred galaxies, and rest- are neutral. In the next interval of time the picture can be alternated. Presumably, matter and energy features on the fabric are also regulated by the principle of least actions. However, there is not enough material for direct observation. The only thing that one can say here is, that cycles' whose length is more than the galactic cycle's length, can be observed only in the systems whose life times are more than a galactic cycle. So, one should reveal like systems spatial and temporal features and detect the cycles, beyond that of the galactic cycles, such as glacial cycles $\approx$ **600 $\pm$50 million** yearly. These are the greatest cycles, accessible on Earth.

12. Task's Mathematical Setting

Let's now formulate the above-mentioned problem from the point of mathematics. The angular momentum usually is defined by **A** = **mvr**, where **m** is mass of body, **v**-its circular velocity, and **r**-the distance between centre of rotation and body. Let's analyse the maximum of the function **A.** If the domain, each of the variables known, then the trivial maximum of A is obvious: $m^*v^*r^* = A_1$ where **m***, **v***, **r*** are right-sided values of the corresponding domains. In what conditions however can A_1 increase? It seems incredible, but it is possible, if one may, uniform movement like **mvr**, replaced, but somehow connected, in regard to the same axis of rotation, differential movements-$m_iv_ir_i$. From this cause let's consider some astrophysical scales and some formulas and results. If the substance' the average density in the space is ρ, then $m = 4/3\pi r^3\ \rho$, (or $m = \pi r^2\rho$, in case of plan) and $v = Gm/r$ (Binney, Tremaine, 1987; Alfven, Arrenius 1976). So, **A** is a function, actually, with only one variable. The distance **(0, R)** we somehow subdivide the **n** pieces, so that $r_i = R_i - R_{i-1}$, $i = 1,2...n$, $\sum r_i = R$. It is not difficult to define the corresponding m_i and v_i for each of the rings. Thereafter, if every ring is able to revolve in a different velocity, then it is possible that the resulting momentum- $\sum m_iv_ir_i = A_2$, in regard to the same axis of rotation, will be greater then A_1. Since in life as well as in death the mass of the star determines its fate, let's consider the partition of variable **m** instead of subdividing variable **R.** Then, the above-mentioned task can be expressed: the mass optimal partition (or quantization) and its differential revolution can increase the total momentum. Apparently, the A_2 increases further, at a certain procedure. Why does accretion take place in space? After all, doesn't the partitioned mass within ring $(R_i - R_{i-1})$ together, already revolve differentially producing additional (secondary) angular momentum and compensation? The accretion, nevertheless, is important. It congregates within ring $(R_i - R_{i-1})$ scattering interstellar dust and clouds so that it later transmits to the mass a gyration around its centre of weight- ω_i.. The gyration (as a small parameter), in order to be reasonable, should have a speed, in an order distinguished in revolutionary speed. In that case only the gyration is significant, and can expand the previous angular momentum A_2. So, we can expect

$$A_3 = \sum(m_i\, v_i\, r_i + m_i\, d^2_i\, \omega_i) > A_2 \qquad (3)$$

Where, d_i is the radius of gyration of the body, ω_i is its the angular velocity (Norris, 1953). That is not the whole though. The interstellar substance's other (tertiary) partition and accumulation around of the earlier bodies, with the corresponding revolution and rotation, produce an additional increase of A_3 (or a decrease of total restoration length) and so on. Why is the quickening of **A** pursued? The reason is obvious: due to the simultaneous differentiated movements' (revolution and rotation) the total angular momentum constantly increases and the balance is restored progressively in shorter time intervals. It is important that the system's reactions should be formed and accomplished faster then the external force is able to change. Also, it is significant for a other reason: the decrease of total restoration length is achieved in a fixed mass' conditions. Otherwise, we get the maximum of total angular momentum for the given **m**. The equivalent task is: the given angular momentum received by the minimum of **m**. The interstellar dust, clouds further partitions and accumulations around the earlier bodies, as was already noted, can produce a new increase of total angular momentum. The mass partition is not infinity though. Its limit is defined by; 1. The coupling means, which provide interchange 2. The energy flow through the system. In such a way, one can define the very massive and the very petty celestial bodies and corresponding cycles, which are true limits of possibilities. Otherwise, one can determine the organized (embedded) bodies' hierarchic system, which realize completely the equilibrium. As noted above, the limits and restrictions stipulated by alteration of a angular momentum, are overcome in diverse ways (substance regular quantization, cycles', such as centre of mass, revolutionary and rotary range, direction, radiation). Undoubtedly, the long-term cycles are not limited only by the offered. There are many specific cycles not spoken of here. For example, binary and double stars' cycles or cycles of pulsars and so on, whose preceding and succeeding cycles are unknown as of yet. Thus, the initial mass' optimum partitions (and, determination of motions' allocation) in space is the result of maximum total angular momentum, and therefore, of restoring equilibrium faster and more completely. Otherwise, the cause of **m** and **v** optimum quantization in space, is a total angular momentum maximum expenditure in the given interval of time. Why is, so much attention consecrated to the third-, forth- and other light rate's angular momentum? Because they can collapse the system, accumulating during a galactic cycle. Therefore, as the compelled angular momentum scrupulously is analyzed and balanced the corresponding system will survive longer. The comets, and Venus' inverse rotations are worthy testimony for such precision.

13. Variation Principle Selection

What could force all of these, to proceed in the indicated order? One should not seek unusual force for the answer. We have only to check which principle could generate orderliness in the course of changing external conditions. Since the changes of energy are continuously optimized, then the space is subjugated to the principle of least actions – **PLA, min** $(mvs = mvds/dt\ \Delta t = mv^2\ \Delta t)$. Hamilton's principle is the only trigger for interstellar substance (and so, motions) optimal partition or quantization in the given cycle. The principle of least actions stipulates the astronomical events' extreme characteristics, and their spatial and temporal organization (co-ordination and subordination). So, any definite size's outer space, can be self-organized under the principle of least actions, regardless of it is empty or not (such as the solar system). It permanently is improved and adapted to the compelling external forces and it equilibrates them in the shortest time. The **PLA** has a grand property, it is applicable persistently everywhere, in all levels of matter organization, without any limitation. In this respect in 1959 Polak, L.S. notes "We do not yet know, why the significant part of physical nature phenomena, known to us, is packed in the variation scheme, and why the significant part of physical science, from the mathematical point of view, is considered as class variation calculation" (Polak, 1959)]. Unfortunately, this situation by noting has changed present times.

14. Task's Formulation by Optimization's Terms

Taking into account the above reasoning, one may under consideration of the problem produce correctly a multistage process, of discrete, deterministic type. The purpose of formulating such a mathematical model is not so much to calculate numbers. But, rather to determine the structure of the solution and to understand characteristics of the systems, which determine the decision (transformation) to be made at any particular stage of the process. So, let's examine the objective (return) function

$$\min_{m_i} \sum_{v=1}^{n} \left| A(T_j) - \sum_{i=1}^{N_v} \varphi_i(m_i, v_i, r_i) \right| \qquad (4)$$

where the $A(T_v)$ is an angular momentum, which is compelled by the galaxy centre to an interstellar dust and clouds (with the total mass $M(T_v)$ on $(R^{**} > R^*)$ where **R** is distance from the galaxy centre), at the time interval T_v, $v=1,2,3,..,n$. The functions φ_i, $i=1,2,3,...,N_v$ are local in regard to the galaxy angular momentum, determined by the expected mass accretion in a pointed area of space. They represent the angular momentum on the unit of planet mass, connected with its revolution, rotation and its satellite movements. The restrictions are $\sum m_i = M(T_v)$ and the current cycles length. Usually, the problem substance's quantization shouldn't to be considered solved in the mathematical sense until the structure of the optimal policy is understood. Otherwise, the mathematical solutions do not reflect solutions of an astrophysical sense. So, let's do some simplification. As $A(T_v)$ is a large parameter and changes slowly at each T_v, then the solution of (4) determines the optimum quantization of M on (R^{**}, R^{*}) for each v, with the common axes of revolving. From the cyclones (or anticyclones) nature in the atmosphere and hydrosphere one may assert that the $A(T_v)$ is an unimodal function. There is not any reliable statistics about $A_v^* = \max A(T_v)$, and T_v^0, T_v'' namely the function domain's left and right points. Therefore, we are able to operate only by functions φ_i, $i=1,2,...,N$. From this cause, instead of (4), let's consider the equivalent objective function

$$f_{N,s}(M) = \max_{m_i} \sum_{i=s}^{N} \varphi_i(m_i, v_i, r_i), \quad 0 \le s < N \qquad (5)$$

in the restrictions 1. $\sum m_i = M$, 2. Prevailing lengths of cycles. The index s emphasises the time-dependants on a process, otherwise we suppose that the functions $\varphi_i(m_i, v_i, r_i)$ are changed over time. Since the functions $\varphi_i(m_i, v_i, r_i)$ are actually one variable function, for the simplicity we will use only one variable. Using the further particularisation of the structure of the criterion function φ_i, $i=1,2,...,N$, instead of the (5), one may write

$$\max_{u,v,w} \sum_{i=s}^{N} \{ \varphi_i(u_i) + \sum_{j=1}^{n_i} [\psi_j(v_j) + \sum_{l=1}^{m_j} g_l(w_l)]\} \qquad (6)$$

or using the obvious properties of the maximum operation and the separable structure of the return function we will receive

$$\max_{u_i \le U} \sum_{i=s}^{N} \{\varphi_i(u_i) + \max_{v_j \le u_i} \sum_{j=1}^{n_i} [\psi_j(v_j) + \max_{w_l \le v_j} \sum_{l=1}^{m_j} g_l(w_l)]\} \qquad (7)$$

the restrictions, are 1. $\sum u_i = U$, $\sum v_j = u_i$, $\sum w_l = v_j$, 2. Prevailing lengths of cycles. In (7) the allowable the decisions set, unlike the classical approaches are not fixed, which causes severe advantages and, at the same time, some discomfort. Further, here the functions φ_i, Ψ_j, g_l present respectively the primary, secondary, and tertiary bodies' revolutionary and rotary angular momentum together. They are of the $(mvr + md^2\omega)$ type. The measurement of variables v and ω, as we have mentioned above, is of different scales

$$mrv(t) + md^2 \sum\omega(\tau) \qquad (8)$$

where $t >> \tau$. As we have already noted, the mean values of $v(t)$ and $\omega(\tau)$ are determined by m_i (exactly by r_i). The determination of v_i and ω_i is basically executed according to the entire com-pensation local (regional, territorial, or global) angular momentum during the corresponding cycle, and to satisfy the condition (1). Therefore, the relations between v_i and ω_l, $i=1,2,...,n$, are coordianative and subordinative. Such movements one may call laminar circulation, in the analogy of laminar currents in hydrodynamics, considering energy's minimum dissipation. Any other movements (such as synchronic, turbulence) originate wastefulness of mass-energy and may be dismissed as "impossible". However, some details can not be understood completely. For example, should each quantization (or differentiation) be controlled from one centre? Or is some decentralization of control necessary? Which organization of the control is optimal? These questions most likely are more of an organization's theory, then - an astronomical or physical question.

15. General Restrictions and Constraints

Because the solar system basically involves three types of variables with different speeds, the balance equation is formed for each speed separately. The general balance equation therefore involves three types of variables: slow, middle, and rapid. The slow and rapid variables are just parameters (called large and small parameters) for the middle variable. They are changed slower and faster with respect to the middle variable. The middle variable, in the other hand, is a large parameter for the next scale balance equation, and at the same time, is the small parameter for the previous scale balance equation. So, we have balance equations in the following system.

$$\frac{du}{dT} = U - a_0 u - \sum_i \alpha_j u v_j \qquad u(0) = a,$$

$$\frac{dv_j}{dt} = \alpha_j u v_j - b_j v_j - \sum_k \beta_{kj} v_j w_k \qquad v_j(0) = c_j \;, j = 1,2,\ldots,n_i \qquad (9)$$

$$\frac{dw_k}{d\tau} = \beta_{kj} v_j w_k - d_k w_k \qquad w_k(0) = l_k, \; k = 1,2,\ldots, m_j$$

were **u, v, w** are slow, middle, and rapid variables respectively. The following are relationships between these variables.

$$u_T, \qquad v_T = k_1 v_t, \qquad w_T = k_1 k_2 w_\tau$$

$$u_t = 1/k_1 u_T, \qquad v_t, \qquad w_t = k_2 w_\tau \qquad (10)$$

$$u_\tau = 1/k_1 k_2 u_T, \qquad v_\tau = 1/k_2 v_t \qquad w_\tau,$$

k_1, k_2 are changed from cycle to cycle. Talking about the balance conditions one should understand the possibility of the excitement conversion- **a, b, d** transmission- α, β and dissipation-**a,b,d** in its totality during each cycle. The verification of the coefficients α, β, **a, b, c, l, d** and their exact values is important. The same issues are related also to the k_1, k_2. The completeness of (9) demands involvement in each equations indirect external influences μ_i, $i > 1$ too.

16. Basic Recurrent Relations

For the optimization problem (7) we are going to use the optimum principle (Bellman, 1959). Let us define $F_s(U)$ as the total return obtained starting with a quantity **U** at stage **s,** and ending at stage **N,** employing an optimal policy, **s = 1,2,...,N.** It is easy to see that the following recurrent relations define solutions of (7).

$$F_s(U) = \max_{u_1} [\varphi_1(u_1) + F_{s-1}(U-u_1)], \qquad F_N(U) = \max_{u_N} \varphi_N(u_N) \qquad (11)$$

$$F_{s-1}(U-u_1) = E_1(u_1) = \max \sum^{n_i} [\psi_j(v_j) + f_j(v_j)] = \max [\psi_1(v_1) + f_1(v_1) + E_2(u_1 - v_1)] \quad (12)$$

$$\{v_j\} \leq u_l \quad {}_{i=1} \qquad\qquad\qquad v_l \leq u_l$$

$$f_j(v_j) = f_{j1}(v_j) = \max_{\{w_k\} \leq v_j} \sum_{l=1}^{m_j} g_l(w_l) = \max_{w_l \leq v_j} \left[g_l(w_l) + f_{j2}(v_j - w_l) \right], \quad j=1,2,\dots, m_i \qquad (13)$$

where **u, v, w** satisfy to the (9) respectively.

What do the solutions (11)-(13) represent? They determine the mass-M_{ijl} , $i = s, s+1,..,N$, $j=1,2,..n_i$, $l=1,2,...m_j$ optimum quantization in the space by size **(R**-R*)**. The recurrent equation (11) determines the mass' primary, the (12) - secondary, and the (13) – tertiary quantiztion. In addition the M_{ijl} , the mean values of circular and angular velocities, and corresponding distants also are determined. Otherwise the recurrent relations (11)-(13) determine the fabric and the velocities distribution, which optimize angular momentum, and its further conversion, transmission, and dissipation, permanently compelled by the galaxy's centre at T_v , $v=1,2,..,n$. The unit of time-**T** is a galactic or any other long length cycle. The solution (11)-(13) were received in the non-stationary condition of galaxy centre's angular momentum, so, the system's structures and functions are formed simultaneously. This means the solar system immediately reacts to any changes of angular momentum by undergoing structural and functional alterations, whose extents, are also determined by the excitement's force. Thus, in the forming of the solar system it is much simpler to use the concepts of a multistage decision (transformation) process than to rely upon analytic manipulation. This concept seems to be a superimposed structure, and is not intrinsic to the natural structures, since they are **self-organized**.

6.6 Self-Organization's Trigger

Let's revise what we did. We were having a process in which we chose of transformation (decision), which could be applied to the system at any time. At each step we chose the value of **m** from a set of allowable vectors of **M**. A vector **m**, was called the decision vector or decision variables and the m_i was called a decision. We determined also the return function. The return function and the optimal policy have a magnificent property: the return function is determined by the optimal policy, and conversely, the optimal policy is determined by the return function. Duality, of this property is also manifested in the other methods of optimization. Our purpose was not so much to determine the return function, but more importantly, to determine the structure of the optimal policy. Simply, we wanted to know the characteristics of the system, which determined the transformation to be made at any particular stage of the process. So, before the process developed we ourselves beforehand prepared all prerequisites and premises. After these clarifications, the system starts to function. That means at each value of external excitement (angular momentum) a transformation (reaction) is selected, which equilibrates corresponding changes. What does the reaction mean precisely? It means the further quantization of **m** when the excitement is above permitted limit, or the **v** and ω cyclic deviations, when they are below a permitted limit. This mechanism of the establishing equilibrium is law. Deviations form, for the chosen **m**, revolutionary and rotary cycles' range. Consequently, further reactions and development of the system are stipulated by the **A(T_v)**, namely by external conditions, which are time-dependent. The increasing (or, on the contrary, the decreasing) systems' structural and functional diversity is a by-product: the main purpose is **min(mvs)**, (in this case **optA(T)** in time and space). And so, evolution is provoked by the alterations of external conditions (energy's sources), which inevitably must be balanced **at the shortest times.** What is more, the improvement (approximation in the space of policy) has its limits, which again are determined by the external conditions. This follows mathematically from dynamic programming.

1. The Principle of Least Actions and Evolution

Did the forming of the solar system occur in the mentioned way? Here, there are two possibilities. First, someone beforehand determines all prerequisites and premises, (time-dependent allowable set of decisions, return function, the principle of optimum, and so on) in which the system later interacts and evolves, as it was executed above. Secondly, there is the universal principle, such as **PLA**, and not any other condition. The prerequisites are gradually formed and expand during the changes of ambient factors. The only requirement is the time-dependence of external conditions, which is common in both of cases. The **PLA** compels that the mass-energy's quantization and conversion constantly have been optimal, if there is unbalanced accumulation (or deficiency). This principle generates the optimal policy (reactions, therefore, the structures also) all the time for every change of the external conditions. They subsequently are preserved and so, an improvement and orderliness emerge. The development as noted earlier is a by-product, the main purpose is **min (mvs).** The assignment (or mission) of all feasible diversities and orderliness, realises that in a more and more effective fashion in the course of changes energy's source. In the **§3** we explicated that the external excitements designate the rising configurations (structures). If the external excitements stereotypically repeat previous values then the structures' preceding forms, and values also are repeated: in other words, the configurations (or responds) remain the same. Oppositely, if the external conditions are changed then new structures arise as a balancing mechanism. As noted, this procedure of establishing equilibrium is law. So, it is impossible to say in advance what will be the organization of structures (and functions) after the several cycles. And so, the assertion that the development's whole course somehow beforehand was programmed, or orderliness (such as brain's, solar system's and the cycles' other stable system present, perfect subordination and co-ordination) was created instantaneously, is vulnerable. This again reverts us back to the question are the structures themselves are only the result of external conditions? So, the answer of the forming of the solar system depends on what lies in the base of evolution. This process (system) evidently is an evolving process. The evolving systems are accompanied by some compulsory stages: origin, infancy, adulthood, old age, and death. These specific events' emphasizing alternations are common. In the sphere of living systems the sequence of like events and their significance was investigated and established by biologists in the XIX century. Their main attributes are selection (reaction) and reproduction due to **PLA**. Starting in the 1950's astronomers, in dealing with evolution of celestial bodies, constellations, and galaxies observed and established the same stages within such objects. In the simplest case, the solar system's adaptations (reactions or selections) represent the mass regular quantizations, the cycles' (centre of mass, revolutionary, rotary) rang, and direction as a response (or a reflection) to the changes of angular momentum. The specific weight of the celestial bodies' life cycles (forming-collapsing) is apparently much greater than is accepted. So, the law of evolution is universal, and it subdues all of those kinds of systems. What is the dependence between **PLA** and the principle of evolution? Is evolution fiction or inevitable? If it is unavoidable, then what triggers it? What triggers evolution is the **PLA**. It is easy to prove following

Theorem: *The principle of least actions is a necessary and satisfactory condition for development and hierarchy of the systems.*

2. Selection as Motivator for Development

In the literature there are some mathematical models reproducing evolutionary systems. The classic examples are equations of Lockta - Volterra, (Svirejev, Logofet, 1978)

$$dx_i / dt = a_i - \sum b_{ij} x_j , \quad i = 1,2,3,...,n \qquad (14)$$

where $x_i > 0$ is the number of specimen type i in the population, a_i, c_i and b_{ij} are coefficients and equations of Fisher- Eigen (Eigen, 1971; Field, 1982)

$$dx_i /dt = E_i - \sum E_i x_i / \sum x_i , \quad i = 1,2,3,...,n \qquad (15)$$

where $E_i > 0$, is dependent on external conditions. The mentioned equations have numerous, modifications, and generalizations. They satisfy theoretical and many practical requirements and reflect various features of the evolutionary systems (processes).

It seems that the both models are vulnerable, since they do not reflect an explicit choosing question. Indeed, in order to execute the selection it is necessary to have an objective (return) function. Because the question is related to the natural systems, the objective function must be universal and applicable for any level of the matter such as cell, tissue, organ, organism, population, species, and biogenesis (the same issues relate also to the comet, satellite, planet, sun, constellation, and galaxies and so on). The objective function should not be artificial or compelled: it should be sought from existing relations or principles. On the base above-mentioned point of view of evolving systems' we can say: what principle is so universal and can provide the improvement. Apparently, it is only the **PLA**. So, determining by (14) and (15) the solutions x_i, $i=1,2,3,\ldots,n$, actually represent the optimal policy, and one tacitly was supposed as an optimization process. In other words, in the evolutionary process only the optimal reactions are selected. This means, the approximation in the space of policy is executed. Therefore, on the base of duality between optimal policy and return function, one can say the return functions represent the real objects. The objects by their reactions (selections) optimize the functional to which they are pointed.

In the theory of optimization (variation calculus, control and adaptive systems) it is well known that any stretch of the extreme curve also is an extreme curve. From the **PLA** it follows that reaction is optimal, directional, deterministic. The **PLA** generates development and self-organization since selections are irreversible. From all of these it follows that evolutionary systems constantly are improved, and develop because all previous reactions are "preserved

3. Stability and Development

The minimisation of energy dissipation takes place constantly due to the **PLA**. There are basically two procedures for the minimisation of energy dissipation: to maximise permanent connections and hold the temporary contacts to a minimum, or on the contrary, to maximise temporary contacts and hold the permanent connections to a minimum (within inner cycles). The first generates stability, the second-development. The development and stability by some sense are opposite processes. The first is associated with the changeable, the second-with unchangeable (all changes were predetermined) external conditions. The development suggests growing and changing, improvement, which are an effect of irreversibility and selection (of reaction). This choice is a result cyclic deviation of the external conditions. As noted, if the external excitements stereotypically repeat previous values, than the reactions' (structures') preceding forms, and values also are repeated. This means, the configurations (or responses) remain the same. Oppositely, if the external conditions are changed, precisely are not repeated, then the new reactions (or structures) arise as a balancing mechanism. The excitement becomes undetectable if its changes are two slow. Thus, the changes of angular momentum (or gravitation) for human being are undetectable because they represent, as noted, spatial structures. There are a spices for those the motionless objects also always are undetectable. Stability means the system's spatial-temporal relations remain invariable in the course of changes' external conditions.

Which way is most preferable for evolution? In the first procedure, in the system beforehand, there is a set of reactions (the instincts - in terminology of biology) for a limited number of changes in external conditions. The system regularly selects the reaction, which corresponds to the definite value of external conditions. The classic dynamic systems basically represent the same relationships. Let's consider the equation of harmonic oscillator

$$\mathbf{d^2x/dt^2 + \omega_1 x = r\cos(\omega t + \alpha)} \qquad (16)$$

which is under influence of the external harmonic force. It is selected priori depending on physical conditions. Here **r** - amplitude of oscillation, α - its initial phase, and a number ω defines frequency of oscillation. The number ω_1 represents own frequency of the oscillator. The general solution of (16), is written by the form

$$\mathbf{x(t) = r_1\cos(\omega_1 t + \alpha_1) + r\cos(\omega t + \beta) / |\omega^2_1 - \omega^2|} \qquad (17)$$

The $\beta = \alpha$, when $\omega_1 > \omega$, and $\beta = -\alpha$, when $\omega_1 < \omega$. The first component of (17) is the solution of corresponding homogeneous equation. Often it is called eigen (own) oscillation of oscillator and describes transition processes. The second component of (17) represents enforced oscillations. It is called also established solution. So, the system in order to be stable reflects all frequencies of the pre-signed external force (Pontriagin,1965). The general solution, repeating the external force's temporary characteristics, realises equilibrium in the shortest time, from here follows the minimum of energy dissipation.

Some bio-systems, such as insects (bees, ants, and termites) were formed in that way. For over **40 million** years these bio-systems have been stable, but have never developed (Chauvin, 1963). Further changes of surroundings remain imperceptibly, undetectable forever, in spite of the fact that, they are perfect and stable systems in regard to selected external conditions. They stay absolutely unchanged, because they've never had premise for advancement. Such species still continue to survive: convert and equilibrate the substance's (or energy's) regular another portion. The external conditions (comestibles' sources), in which the insects are immersed or embedded are still preserved and able to act. If these external conditions are exhausted, terminated or are suddenly changed, the systems will also vanish from evolution. As many other spices have been vanishing (even the species, that appeared in the evolutionary process after insects), when the external conditions, in which they have survived, were sharply changed. Such system was unable create new actions (or structures), and so, noting is accumulated in the course of time. The insects' example is an instructive example in that the cycles' perfect system, such as the human brain, can also be in an analogous situation. Namely, there are many specific cyclic alterations around human brain, which forever had been the imperceptible and undetectable. And so, due to the **PLA**, compared with the human brain, the cycles' more perfect systems can be formed and establish equilibrium around such cyclic alterations. Is it possible, when some of perfect cycles' systems, such as human being, be under influence coursing very fast, like to the radiation, cycles'?

The development's cardinal prerequisite is; to create the temporary contacts within inner cycles (restoration processes). Why? Because the temporary contacts between inner cycles allow the exchange of a cycles information. As a result, equilibrium is established more quickly and perfectly. Instead of balance by substance exchange, the corresponding information is simply exchanged and optimized, so the equilibrium is established faster. What triggers the emerging temporary connections within the inner cycles, and in general, within any type of cycles, is the same **PLA**. These phenomena are prominent and well known within economical cycles. The functions of the economic systems become much more effective and flexible, when instead of substances, the information, or the another equivalents, about them start to circulate. Moreover, the information systems' conception and processes associated with them the computer system, have been formed directly as a result of them. They both operate only by the circulation of information and constantly save time and therefore furnish more freedom: the circulation of substance becomes secondary. Such systems are acquired despite matter's, energy's large expense, however these letter quickly are compensated. The information system's dependence on the external conditions is now determined not only, by the excitement's physical characteristics but the systemic significance that the excitement acquires for the system, during its development (construction, formation, and projection). So, the factor of time is fundamental for the development.

4. Development's Two Equivalent Paths

Undoubtedly the most cardinal achievement of the mentioned replacement is the emergence of temporary connections (the conditional reflexes– by terminology of neurophysio-logy) within the electrophysiological (brain) rhythms. This concept creates "unlimited" progress, novelty and diversity. Systems, with such quality, distinctly extend interacting cycles' spectrum, and equilibrate them to some extent. So, the **PLA** permanently forces that the division (bifurcation) of labor or the temporary connections within the inner cycles have been executed. Since they both, as noted, furnish the equilibrium with minimum dissipation of energy and it is more important the reaction (equilibrium) is restored progressively in shorter time intervals. Again, for the quickening what is needed? The system's reactions should be formed and accomplished faster then the external force is able to change. Apparently, the minimum dissipation of energy and the establishment of equilibrium in the shortest time occur also in reverse to the bifurcation, processes. Particularly, the combination, consolidation or centralization of labor, also furnishes the same effect. In such a way one can observe the ways of development in the biology (different types of symbioses), in economy (centralization of companies, banks, unions), and in astronomy (captures of celestial bodies, cannibalism of galaxies). So, both ways of development are equivalent for evolution. Identical purposes are persuaded in both cases; to optimize allocation and motion of mass-energy (or their equivalents') in formed conditions. In the first case, the whole, performing only a general function, due to the **PLA** is divided continuously into parts by the more and more specific functions. Separate parts get their original function, and become somehow autonomous. This is development by differentiation (or bifurcation). In the second case, again to the **PLA,** an unite combining with other unite get general function and interact with more powerful cyclic alterations. This is development by integration. After the labor's division the system starts to interact with new excitements (energy new sources) in regard to which, the system before was indifferent. In such way the range of excitements increases: the system interacts to more and more diverse sources of energy and "controls" them. In this case the new actions (structures) permanently creates which, as noted, are conserved. Something valuable is created whose disregard will be unreasonable, since it takes on more value then the surroundings. In both cases due to the **PLA** temporal and spatial (functional and structural) relations are constantly restored and preserved. The system is improved and the orderliness gradually increases. For that reason, the second procedure-minimization of energy dissipation is creative and perspective. The question arises, if the emergence of temporal connections were creative, why in the beginning did it not prevail everywhere. Because the all possibilities have an equal chance and no single one has the privilege, in regard to the others, and so, each of them is subject to inspection. This can appear also in the arena of evolution letter. The ranges of cycles become basic mechanisms for the evolutionary systems. For each groups' of cycle the range of fluctuations is deferent. Thus, flora and fauna cycles lengths are changed respectively, by **40%-45%** and **25%-30%** of their the mean values during of changes of external conditions. The cycles of the atmosphere, hydrosphere, and lithosphere as well as Earth's orbital parameter's everywhere fluctuate near their mean values. The famous galactic cycle range involves in **18%** of its mean value, etc. Now, one can say, what type's cyclic alterations are formed in the beginning (in the originating spiral structures).

5. Reactions' Preservation and Accumulation

The laminar movements (currents) in the hydrodynamic are noteworthy in that the energy dissipation is a minimum, since the duration of restoration processes are commensurable with the temporary scale of hydrodynamic phenomena. In non-stationary currents, the liquid, which is more outstanding, "remember" prehistory, i.e. the changes of velocities' field in previous moments of time (in former relaxation processes). The heritage factor becomes essential when the duration of a relaxation process, which has occurred in the medium, is commensurable with the temporary scale of a hydrodynamic phenomena. The heritage factor's role is determined by the Debre's number $\mathbf{De} = \lambda/t_i$, where the λ is the time of the slowest relaxation processes, t_i –the temporary scale of a hydrodynamic phenomena (Monin, Yaglom 1965; Shulman, 1986). So, when the duration of restoration becomes comparable with the duration of an external excitement, the excitement, can somehow be remembered (preserved, registered) in the system. Further, the time of reaction is the time of established equilibrium. Therefore, such an agent provides and forms the reaction sooner then the external forces are able, to be changed. The duration of both is exactly the same, then one could say the system operate by its maximum rhythms. Generally, the reactions are not uniform or homogeneous processes. They represent a multistage process, with particularly consists of 4-5 compulsory stages, that have diverse duration and alternate with one another in some order. This is general and is applicable for any type of system and restoration processes. So, if a system's restoration processes are able to reflect (or repeat) an excitement's temporary characteristics then the energy dissipation will be a minimum, and in addition to it, the excitement can somehow be remembered in the system. Below we will persuade that in the reality noting is conserved, accumulated or remembered in the system. Simply, within restoration processes such connections emerge and are established which furnish and form a reaction by corresponding duration and, it can trigger at the necessity. The last has great significance for the theory of memory. Therefore, the restoration processes' composition or construction is very important.

6. Restoration Processes' Organization

From the cycles organization it follows that the restoration processes (inner cycles) formation depends on the excitements. Particularly the excitement's parameters: magnitude, intensity, frequency, length stipulate the types of reaction. So, the excitement itself provokes the form at in of either duration's restoration processes, or the combining corresponding reaction. How? Due to the **PLA**, the temporary connections, among the restoration processes (structures) emerge, which provide a reaction with comparable duration. This, in turn, requires lastingly, the higher frequency cycles' arrangement in time, and so on. So, the labour's division modifies and subjugates inner cycles continually, and furnishes certain inner structure conforming organization. Selected in this way the sequence of restoration processes determine; what will be the energy dissipation minimum or what will be the remembered, preserved, and registered excitement? That is the reason of cycles' lengths fluctuation. When a necessity arises to remember (or react) to an excitement the relaxation processes are rearranged, reformed, and reorganized. As a result, a new cycle (cyclic change) is created and situated among former restoration processes; reproducing the last excitement's temporary characteristics. So, the excitement constantly forms the sequence of restoration processes (reaction) whose duration has been commensurable with the excitements. Just as the inner medium (cycles) was regulated and ordered excitement's transition and balance constitution will be in time, therefore the energy dissipation will be a minimum. Such cycles' organization has a significant advantage, since it is greatly conducted by searching, forming, and reproducing optimal form interactions with the excitement of a different mod. The same issues, actually, lie on the base of the learning process, the formation or establishment of conditional reflexes, and some other processes.

7. Stochasticity against the Hierarchy

As noted earlier, the solutions, decisions, or reactions are a multistage process. The solutions are associated ordinary with the external forces or conditions, which were given before-hand, whereas the reactions are related to the systems where the external forces are not given in advance: the external force can only be known at certain times: it is cyclic. In like conditions, proceeding from the supposition that the system's interactions take place in uncertainty, sometimes such systems are identified as stochastic (adaptive) systems. Indeed, it seems that there is no reason to concentrate attention on the cycles since they are also changed during these times. Hence one may operate the theory of probability or the adaptive system, and overcome any emerging uncertainties Therefore, the system's behaviours describe and evaluate stochastically, and adaptability. Although, this is not the case, since executing an average among a different scale's cyclic alteration, we actually neglect the cycles' hierarchy, which furnishes a reaction of the multilevel system with the energy' minimal dissipation. Ignoring the cycles' permanent origin as a result of the labour's division, we return to the starting point: how to get a reaction (equilibrium) with minimum energy dissipation under **PLA**. So, instead of developed stochastic methods, one should concentrate the attention on small and large parameters in the theory differential equations, since they more adequate describe the features of such systems.

8. What is the Optimized in the Reality?

Thanks to the dynamics of inner cycles (restoration processes), the system is capable of actively withstanding various excitements. As a law, in the beginning, reactions are directed only to the balance and expending excitement's influence. Then, due to the **PLA**, they gradually are aimed to the support and conservation of the system's unity and created inner structures (functions). Because at this time, something valuable is created whose disregard will be unreasonable, since it makes the energy's conversion more effective, then the surroundings. Therefore, formed later reactions take into account both: excitement and all former achievements. This means, reaction equilibrates not only the excitement but also supports and conserves the unity and integrity of the system. That adds to the improvement and development and makes the equilibrium more persistent. Once created further it never withdraws; it is adjusted, modified and preserved constantly during evolution. In this regard Orbely notes "Regulation of new forms acquired in the course of the evolution are shields, changes, modifies the preceding forms which still preserve their significance even though their function is often is disguised and camouflaged in various extent" (Orbely, 1961). With this regard Aristotle has noted still two thousand years ago, that if something optimum is created in the evolution, then it is preserved further in the evolution. Therefore one may consider newly acquired cycles due to the labour's division, as more complicated and important in regard to earlier existence that are taken (get into) in subordinate situation to the newest. In any event a certain subordination and hierarchy is established between cyclic alterations, collaborating with the realization of these functions (system of reactions).

Therefore, due to the PLA, evolutionary systems, interacting with the excitements of their surrounding, establish equilibrium not only among them, but also conserve former, own all achievements.

9. Cycles as Balancing Mechanisms

As it was noted, the equilibrium with respect to any excitement requires certain time in order to a finish inner transformations and re-constructions. That is why a system, due to effect of excitement of different frequencies, may be balanced with respect to a parameter and be unbalanced with respect to another parameter. And from this fact permanently occurring phenomena of the essential unbalance of processes running in systems, formation and self-organization follow naturally. Open systems have not explicit state function that would have extremal properties. The only extremal property of these systems, apparently is permanent co-ordinating and matching of reconstruction and redistribution during restoration processes. Otherwise it is impossible to carry out processes in due time and in due place. So, in order to co-ordinate, it is necessary that system's parameters would not be fixed, on the contrary, they should be mobile, liable and distributed within a range of finite values. Otherwise, a co-ordination would be reduced to simple equalisation and correspondence of conditions. The following definitions are important for the open (developing) systems:

Difinition1 . *A measure of equilibrium for open, dissipate systems is a limit of parameter changes which still allows a system to maintain optimal interaction with a medium, i.e. to transform external excitement with energy minimum dissipation, without any structural change.*

The limits of the mentioned parameter changes are separated and programmed in the sense that they should include all possible limiting changes of angular momentum. They were not formed at once, but gradually in the course of evolution in respond to development and increase of galactic centre loading. Sometimes these limits seem so usual, ordinary, that their explanations are as if senseless. Meanwhile the study of causes of parameter changes is important because they give rise to regular changes of other orders and natures. Thus, orbital parameters' changes bring forth effects on rotation motion of a system and a regular angular momentum's changes. These latter bring forth, in their turn, redistribution and changes with corresponding periods of tectonic strains in the Earth's crust, which then manifest themselves in changes of magmatism, earthquakes, sedimentation, folding, parameters of oceanic and atmospheric turnovers and currents, i.e. their volumes, velocities, periodical oscillations of axes, and so on.

What is the role of the solar systems matter (masses) with respect to external medium from which it was born and separated? If the angular momentum, forced by the galaxy's centre, begin to exceed the solar system's dissipate capabilities, (otherwise permissible limits of orbital, rotational and other dynamic parameter changes) , then balance should inevitably give rise to new accretion and compression of matter which should bear a job for balance and dissipation of this additional angular momentum. Hence, matter's an accretion in the solar systems, which were separated from medium, always was balancing and dissipating agents for extra force caused by galaxy centre loading. From this it follows that matter and energy in the cosmic space (the solar system) are not infinite and dissipated accidentally. There every portion is strictly controlled and balanced, quantized (stratified). The existing mass' quantization reflects regularities of changes in angular momentum in the course of time.

So, the excitement begins to exceed quantitatively a measure of system's equilibrium, if possible shifts of the system's parameters are already unable to carry out transformation, dissipation processes, which are arisen due to excitement. In other words, if cyclic changes are unable to realise transformation of given volume in a given time interval then a necessary to balance interaction give rise to labour's division, irreversible changes.

Difinition2 *.The coast of equilibrium is the emerging of non-cyclic, irreversible changes, when due to the PLA frequency doubling's phenomenon takes place. In the result of labor's division, such inner cycles and structures are formed, which transform external loading with energy minimum dissipation.*

Thus, cycles represent the balance and stability forming mechanisms. The basic direction of their development is more focusing on co-ordination and regulation. This tendency is accompanied functions' (and so, structures') quantization and stratification, and their deepening. In the course of changes external conditions' they become strong protective means and structures, which counter-respond and balance changes in external selective condition. They form an inner organization of a. system, and determine its level. In the definition level of an organization it seems, preferable to use the concept of energy quality, instead of cycles' range. But, the system reacts not only to substance and energy changes but also information, signals, and others that are cyclic in nature too. So, in order to consider and involve all of the set excitements, it is convenient to use cycles and their compositions. Therefore, the level of organization is determined by the inner cycles' combination. The organization's level becomes higher as its cycles' composition gets wider: the high level's organization "controls" numerous excitements and establishes equilibrium among them by energy minimum dissipation. The coupling means restrict the level of the organization.

10. Direction or Irreversibility of Energy Transformation

The energy is transformed constantly at the interactions. As was noted, it is not cyclic or reversible. It, on the contrary, is irreversible and has definite direction. So that, the energy quality seems to reduce gradually, but not totally. In the interactions the balance's equations have basically two types of terms: non-linear and linear. The non-linear terms represent the energy transformations and transitions, which the system executes inevitably. These terms are formed by values of energy inflow and outflow in each equation (frequency, cycle). The linear terms represent energy expenditures directed to the preservation system's structural and functional relations. Otherwise, they are necessary for the realisation or execution energy's mentioned transformations and transitions, which during the time form system's development. The development is accompanied by the emerging of high level organization (substance, system). It means, due to the PLA, labour's division (bifurcation) takes place permanently. As a result, such reactions (restoration processes, connections) are created and formed among system's inner cycles, which provides the energy minimum dissipation. Due to the PLA again, these reactions are accumulated and preserved constantly. So, during the interactions the systems are improved, their interaction sphere increases; simply they establish equilibrium among excitements which number increases more and more. These all require energy's inevitable expenses; nothing can be created and maintained without of the energy's expense. This is a law, and it is displayed everywhere. Thus, the development's first step naturally is the emerging of heavy chemical elements and their combinations in the Earth. The development's second step is the forming of the granites[1] in the lithosphere as a final stage of the rocks' transformation, and increasing their supplies at the geologic axis of time. The development's next step in the limits of Earth is the origin and expansion of species and their improvement in the biosphere and so on. The energy's minimum dissipation has compelled first of all the forming and improvement of new structures (the emerging labour's division or bifurcation) and cycles in the our galaxy continually. Moreover, the dynamic distribution of energy (tension) on the cellular fabric[2], according to the cycles' organization, is stipulated by the external, in regard to considering system, excitement and transfers, equilibrates it (the excitement). The irreversibility of energy transformation enforces to suppose the existence more and more slow spreading energy's sources, which are transformed repeatedly with minimum dissipation and creates a high level organization and orderliness by huge scales. This brings us again to the infinity embedded medium's idea. How substantial it is, further research will show.

1. The granites (in general, graniteoits) are noteworthy among other rocks by them, that they are final ring of substance's conversion (metamorphosis, metasomatisis) chain in the crust of Earth. The granites after formation remain unchangeable and never are subjected to the metamorphosis again. Their chemical composition and texture are no longer change, whatever the surrounding conditions be they: pressure, temperature, acidyl, aqueous (watery, hydrous), pH, Eh, and other conditions. The condensed energy in the granites is higher then the condensed energy in the other rocks. (Personal announcement of Dr. Guyumjyan)

2. .In nodes of which the ordinary or barred galaxies are formed, develop (Sec. 6.3- 6.7,).

REFFERENCES

To **Chapter I**

1. Alphven, H., Arrhenius G. *Structure and Evolution History of the Solar System.* San Diego, California, 1976, 331pp.
2. Dubovshin, G.N. *Celestial Mechanics.* Moscow, Nauka,1983,370 pp.
3. *Instabilities in Dynamic Systems.* Moscow, Mir, 1982, 210 pp.
4. Ivanenko, D.D. *On Rotation of the Universe, in Problems of Theory of Gravitation and Elementary Particles*, 17. Moscow, Nauka, 1986, 172 pp.
5. Molchanov, A. M. *On the Resonance Structure of the Solar System*, in: Modern Problems of Mechanics and Astrodinamics. Moscow, Nauka, 1973, 250 pp.
6 Pibls, P., *Structure of the Universe in Large Scales.* Moscow, Mir, 1983, 305 pp.
7. Szebehely, V. *Theory of Orbits,* Moscow, Mir, 1982, 358 pp.
8. Eigenson, M. S. *On the Analogy Between the Solar System Structure and the Galaxy Structure.* Astronom. Zhurnal, No.1, 1963.
9. *Tides and Resonance in the Solar System.* Moscow, Mir, 1975,287 pp.
10. Beatty, J., and Chaikin, A., eds. *The New Solar System.* 3rd ed. Sky Publishing and Cambridge University Press, 1990.
11. Morrison, D., and Owen, T. *The Planetary System.* Addison Wesley, 1988.

To **Chapter II**

1 Vitniskii Yu. M. *The Solar Activity.* Moscow, Nauka, 1983, 190 pp.
2. Gledzer E. B. *System of Hydrodynamic Type and Their Applications*, Moscow, Nauka, 1981, 366 pp.
3. Cox J. *Theory of Stellar Pulsation.* Moscow, Mir,1983, 368 pp.
4. New Cark K., Freizner K. Cycle of Solar Activity, in: *Fizika za Roubezhom,* Moscow, Mir, 1983,204 pp.
5. Prist E. G.*Solar Magnetohydrodynamics*, Moscow, Mir,1985,300 pp.
6. Severnii A.B. Global Oscillation and the Internal Structure of the Sun. *Vestnik AN SSSR*,1981, No. 11.
7. *Hydrodynamic Instability and Transition to Turbulence.* Eds H. Suinin, J. Gallaba, New York, Springer,1981,325pp.
8. Walker A. G. Golden Age of the Sun Physics in: *Fizika za Roubezhoom,* Moscow, Mir,1984, p 159.
9. Ebeling V. *Structure Formation in Irreversible Processes.* Moscow, Mir, 1979, 320 pp.
10. *Physics of Solar Activity.* M, Nauka, 1983, 203 pp.
11. Eigenson M. S. *Sketches of Physical-Geographical Manifestation of Solar Activity.* Lvov, Lvov University Pubishing House, 1957, 230 pp.
12.Foukal, P., "The Variable Sun." Scientific America, February 1990.
13. Wallenhorst, S. "Sunspot Numbers and Solar Cycles." *Sky &Telescope*, September 1982.
14. Wolfson, R . "The Active Solar Corona." *Scientific America*, February 1983

To **Chapter III**

1 Avdeev, A. V. On the Problem of Galactic Geochronology. Publishing House of Academy of Science of Kazakh SSR, ser. Geology, 1973, No 2.

2 Geodynamics, Moscow, Nauka, 1979, 210 pp.

3 Balukhovskii, N. F. *Geological Cycles.* Kiev, Naukova Dumka, 1966, 167 pp.

4 Belousov, V. V. *Fundaments of Geotectonics*, Moscow, Nedra, 1975, 367 pp.

5 Bott, M. *Internal Structure of the Earth,* Moscow, Mir, 1979, 307 pp.

6 Vegener, A. *Origin of Continents and Oceans.* Moscow, Mir, 1984, 270 pp.

7 Gudenberg, B. *Physics of Entrails of the Earth.* Moscow, 1963, 269 pp.

8 Dech, V. N., Knoring, A. D. *Method for Study of Periodical of Ocean.* Moscow, Nedra, 1981, 272 pp.

9. Calibina, I.V, Katterfeld, C.N. The Types and Systems of Planets' Lineaments. ***Proceedings AN SSSR, Seria Geologicheskay, # 11,1975***

10 Zharikov, V. I. *Internal Structures of the Earth and Planets.* Moacow, Nauka, 1983, 410 pp.

11 Kosigin, Yu. A., Votakh, O. A. Hierarchy of Geological Objects and Tectonics. *Dokladi AN SSSR*, 1972, v. 207, No 2.

12 Kropotkin,. A Possible Role of Cosmic Factors in Geotectonics. *Geotektonika,* 1972, No 2

13 Krut, N. V. *Introduction to the General History of the Earth.* Moscow, Misl,, 1978, 366 pp.

14 Kunin, N. Ya., Sardonnikov N. M. *Global Cycle Recurrence of Tectonic Motions* Byull. MOIP, ser. Geology, 1076, No. 3, pp. 3-26.

15 Magnistskii, V. A. *Internal Structure and Physics of the Earth.* Moscow, Nedra, 1965, 379 pp.

16 Onoprienko, V. I. *Problem of Cycle Recurrence in Theoretical Geology.* Geolog. Zhournal, 1976, No. 6, pp. 56-64

17 *Tidal Deformations of the Earth*, Moscow, Nauka, 1975, 186 pp.

18 *Stability of Geosystems.* Moscow, Nauka, 1983, 88 pp.

19 Sadovskii M. A. Automodelity of Geodynamic Processes. *Vestnik AN SSSR*, Moscow,1986, No. 8

20 Safaryan N. E. On Some Characteristics of Geological Cycle Recurrence, in: *Mathematical Methods of Cycle Recurrence Analysis in Geology.* Moscow, Nauka, 1985, 147 pp.

21 Khain V. E., Mikhailov A. E. *General Geotectonics*, Moscow, Nedra, 1985, 326 pp.

22 Tamrazian G. P. Some Principal Planetar Tectonic Regularities and Their Casual Relationships. Izv. Vuzov. Geologiya Razvedka, 1967, No 11, pp.3-17.

23 Hallem E. *Great Geological Disputes.* Moscow, Mir, 1985, 216 pp.

24 Safaryan N. E. Geological Cycles. *Gitutyun ev Tekhnika*, Yerevan, Nos. 11-12, 1987, (in Armenian).

To **Section 3. 2**

1. Alfven, H., Arrhenius, G. *The Evolution of the Solar System.* San Diego, California, 1976.

2. Barenblatt, T.I. *Similarity, Automodelity, Intermediate Asimtotika,* L Gidrometoiztat, 1982. (in Russian).

3 Belousov, V.V. *The Basics of the Geotectonics*, M. 1975. (in Russian).

4. Bott, M., *The Interior of the Earth, its Structure, Convection and Evolution*, 2-ed ed. Edward rnold, London, 1982.
5. Bullen, K.E., *Earth's Density*. London, Chapman and Hall, 1975.
6. Burkov, V.N., *The Basics of the Mathematical Theory Active Systems*, M. Nauka, 1974. (in Russian).
7. Cook, A.H., *The Dynamical Properities and Internal Structures of the Earth, the Moon and the Planetes,* Proc.Roy.Soc. London,A328,301-336,1972.
8. Demenistkaya, R.M., *Ocean Natural Physical Field*. M. Nedra, 1981. (in Russian)
9. Dirac, P.A.M., Long Rang Forces and Broken Symmetric. *Proc.Roy.Soc.* London, A333, Roy. Soc.London, A333,403-418,1973.
10. Dirac, P.A.M., Evolutionary Cosmology. Commentaries *Pontif. Acad.Sci.* 2, No46, 1972.
11. *Earth's Crust and Upper Mantle*, M. Nauka, 1972. (in Russian).
12. Elasser, W.M., *Causes of Motions in the Earth's Core.* Report, ONR,project N-ORI.Inst. Advaneced Study, Princetion 1950.
13. Ferhugen, J.,and others. *Introduction to the General Geology.* v. I, II, 1972.
14. *Geodynamics: Applications of Continuum Physics to Geological Problems.* Donald L. Turcothe, Gerald Schubert, 1982.
15. Fank-Kamenestky, D.A., Physics' Laws an Geological Cycles, *Priroda,* 1964,1. (in Russian).
16. *Gravitation and Topology*, M. Nauka, 1966. (in Russian).
17 Gutenberg, B., *Physics of the Earth's Interior*. New York, Academic Press, 1959,
18. Haken, G., *Advanced Synergetics; Instability Hierarchy of Self-Organizating Systems and Devices*. Berlin: Springer Verlag,1983.
19. Hayashi, Ch., *Ann.Astr.Astrophys.* 1966,v.4 p.171.
20. Hlobustov, A.A., Manner of Natural Interactions. *Zapiski Uzbek.Otdela VMO,* 1971,vip.24. (in Russian).
21. *Hydrodynamic Instability and Transition to Turbulence.* Eds. H Suinin, and J.Gallaba, New York: Springer,1981.
22. Jeffrey, W., *The Earth,* 4th ed. Macmillan, New York,1959.
23. Khain, B.E., *The General Geotectonica.* M. Nauka,1973. (in Russian).
24. Khain, B.E., *Orientation, Cyclicity and Nonuniformity of the Development Earth's Crust.* In "Structure and Development Earth's Crust" M.Nauka, 1964. (in Russian).
25. Kosigin, U.A., Votach, O.A. Hierarchy Geological Objects and Tectonics, *Dokl. AN USSR*,1972, v.207, 2.(in Russian).
26. Kropotkin, P.N., The Possible Role Cosmic Factors in Geotectonics. "*Geotectonica*", No 2,1970. (in Russian).
27. Landscheidt, T. Cosmic Cycle Analysis, *Cycles,* 1995, 45:4 June/ July
28. Lorenz, E. N., *Nature and Theory of the General Circulation of the Atmosphere*. Geneva: World Meteorlogical Organization. 1967.
29. Loustich, E.N., Convection in the Shells of the Earth, According to the Pikeris' Calculations.

Izv. AN.SSSR, ser .Geophysica, No 5. 604,1957. (in Russian).
30. Magnistky, V.A., *The Inner Structure and Physics of the Earth*. M. Nauka, 1965. (in Russian).
31. Munk, Y., and G. Macdonald, *The Rotation of the Earth*, Cambridge, University Press, 1960.
32. Odinstov, M.M., *Geological Sciences, Perspectives and Problems,* Nov.Nauka,1980.(in Russian).
33. Oort, A.H., On Estimates of the Atmospheric Energy Cycle. *Monthly Weather Review v.* .92, No 11. November,1964.
34. Orlenok, V.V., *Physical Basics Evolution of Perispheres*, L.Nauka,1980. (in Russian).
35. Rossby, C.G., On the Vertical and Horizontal Concentration of Momentum in Air and Ocean Currents, *Teblus,* 1951,v.3, No1,p. 15-27.
36. Rossby, C.G., A *Comparison of Current Patterns in the Atmosphere and in the Ocean* Basins.IUGG General Assembly Brussels,1951.
37. Runcorn, S.K., The Earth's Core, *Trans. Amer.Geophys.Union*,35,49-52,63-78, 1954.
38. Sadovsky, M.A., Automodelity of the Geodynamical Processes, *Vestnik AN.USSR,* 1986,8. (in Russian).
39. Sarafyan, N.E., *Some Aspects of Self- Organizatiuon.* (in press).
40. Sarafyan, N.E., *Geological Cycles*, Yerevan 1990.(in Armenian).
41 .Sarafyan, N.E., Movement of the Excitement in Non-homogenous by Relaxation Media. *Izv. AN Arm. SSR, ser.Nauki o Zemle,*No 12,1984. (in Russian).
42. Sarafyan, N.E., Stability and Stabilized Mechanisms of the Earth. *Izv.AN.Arm SSR,ser .Nauki o Zemle,* No 6,1985. (in Russian).
43. Sholpo, V.N., *Earth's Structure: Orderliness or Disorderliness*, M.Nauka,1980. (in Russian).
44 Starr, V.P., *Physics of Phenomenon with Negative Viscousity,* New York,McGraw- Hill,1968.
45. Sverejev, U.M. and D. Locofet, *Stability Biological Associations*. M.Nauka,1978.(in Russian).
46. Tamrazyan G.T. The Greatist Planetic Tectonic Laws and Theirs Causal Links. *Iz.Vyz. Geology and Exploring (Geology)* #11. (in Russian)
47. *The Solid Earth: An Introduction to Global Geophysics*. C.M.R.Fawler, Cambridge University Press, 1990.
48. *Tides and Resonance in the Solar System.*, Nauka, M. 1975
49. *The Instability in the Dynamic Systems*. Ed. Szebehely V., M. Mir,1983. (in Russian).
50. Urey, H.C., *The Planets*, Yale Univ.Perss, New Haven, Connecticut,1952.
51. Vinogradov, M.B., and others. *Theory of Waves.* M. Nauka,1979.(in Russian).
52. Volterra, V., *Lesons sur la Theorie Mathematique de la Lutte Pour La Vie Gauther,* Paris: 1931.

To **Section 3.3**

1. Anarin V.I., Kiselev V. M., Goshko E. Yu. Megacycle Distinguishing by Means of Spectral Analysis, in: *Mathematical Methods of Cycle Recurrence Analysis in Geology.* Moscow, Nauka, 1984, pp. 53-57.

2 Korogodin Yu. N., Leontiev V. I., Baukrev G. F. Investigation of Rock-Layered Associations of Neogene Molasses, in: *Mathematical Methods of Cycle Recurrence Analysis in Geology*. Moscow, Nauka, 1984, pp. 6-26.

3 Sadoyan A.A.Pecularities in Flisch Formation of Danian Eocene of the Armenian SSR and Contiguous Regions. *Litologiya i Poleznie Iskopaemie*,, 1983, No. 2. pp. 43-56

4 Safaryan N. E. On Nature and Mathematical Methods of Investigation of Geological Cycle Recurrence, in: *Mathematical Methods of Cycle Recurrence Analysis in Geology*. Moscow, Nauka, 1984, pp. 171-176

5 Harbrook J., Bonhen-Carter G. *Simulation in Computer in Geology*. Moscow, Mir, 1974, p 319.

To **Chapter IV**

1. Brayan K., Sarmiento J. Simulation of Ocean Circulation, in: *Climate Dynamics*, Leningrad, Gidrometeoizdat, 1988, pp. 430-454.
2. Burkov V. A. *General Circulation of the World Ocean*. Leningrad, Gydrometeoizdat, 1980, p. 247.
3. Gavzilov. *Phenomenological Structures of the Earth*. Moscow, Nauka,1978, p. 144.
4. *Hydrodynamic Instability and Transition to Turbulence*. Eds. H Suinin and J.Callaba, New York Springer,1981, 317 pp.
5. Dragan Ya. P. et al. *Probabilistic Analysis Methods of Rhythmics of Oceanic Processes*. Leningrad, Gidrometeoizdat, 1987, p. 317
6. Investigations of Tsunani, No. 3., 1988
7. *Changeability of Physical Fields in the Atmosphere over the Ocean*. Moscow, Nauka, 1983, p. 168.
8. Kagan B. A. Modern Situation of Tidal Energy Dissipation Problem, in: *Issnes of Ocean Dynamics*. Leningrad, Gidrometeoizdat, 1984, p. 79.
9. Lebedev, V. A. et al. *Ocean as a Dynamic System*. Leningrad, Gidrometeizdat, 1976, p. 203.
10. Malcus, G. Large-Scale Interaction, in: *The Sea*. Leningrad, Gidrometeizdat, 1965, p. 153
11. MaksimovI.V.*Geophysical Forces and Waters of the Ocean*. Leningrad, Gidrometeizdat, 1970, p.456
12 Menabe, S., Brayan, K. *Climate and Ocean Circulation*. Leningrad, Gidrometeizdat, 1972, p. 192
13. *Simulation of Hydrophysical Fields and Processes in the Ocean*. Moscow, Nauka, 1986, p. 215
14. Monin, A. S. et al. *Changeability of the World Ocean*. Leningrad, Gidrometeizdat, 1974, p. 262
15. Sea Currents, *Soviet Big Encyclopedia*, vol, p. 592
16. Ocean, *Soviet Big Encyclopedia*, v. 18, p. 320
17. Oort, A. H., "Balance Conditions in the Earth Climate System." In: *Climate Dynamics*. Leningrad, Gidrometeizdat, 1988, p.91
18. Pertsev, B. P., Ivanova, M. V. "Horizontal Displacement of the Earth Surface under the Influence of Sea Tide Loading.". I: *Simulation of the Earth Deformation and its Rotation*. Moscow, Radio i Svyaz, 1985, pp. 5-10
19. Plakhotnik. A. F. *Changeability of Physical Phenomena and Ocean Properties* . Moscow, Nauka, 1984 , p. 342.
20. Sarkissyan A.S.*Numerical Analysis and Forecost of Sea Currents*. Leningrad, Gidrometeizdat, 1977, p. 180

21. Stommel, G."Review of Sea Current Theory." In: *Problems of Oceanic Circulation* Moscow, Mir, 1965, p. 312
22. Fedorov, K. N. *Physical Nature and Structure of Oceanic Fronts.* Leningrad, Gidrometeizdat, 1983, p. 296
23. Feodorov, K. N. "On Evolution of Whirls in the Ocean.' In: *Issues of Ocean Dynamics*, Leningrad, Gidrometeizdat, 1984 p. 128
24. Felzenbaum, A. I."Dynamics of Sea Currents." In: *Hydromechanics*, Moscow, VINITI, 1970, p. 339.
25. Filander, S. J."Tropical Oceanography." In: *Climate Dynamics*, Leningrad, Gidrometeizdat, 1988, pp. 456-471
26. Folanov, I. P. "Dynamics of Oceanic Currents." In: *The Sea.* Leningrad, Gidrometeizdat, 1965, p. 463
27. Oort, A. H. "Balance Condition in the Earth Climate System." , *Adv.Geophys.* 1985, v l28A,pp. 75-78
28. Oort, A.H.,Peixoto, J.P. "Global Anagular Moment and Energy Balance Requirements from Observations." *Adv. Geophys.* 1983, vol. 25, pp. 355-490.

To **Chapter V**

1. Golitsin, G.S. *Introduction to Planetary Atmoshere Dynamics.*Leningrad, GidrometIzdat,1973, p.104
2. Maderin, V. S. "A Model of Large-Scale Interaction of the Ocean and the Atmosphere.".In: *Issues of Ocean Dynamic Whirls*, Moscow, Mir, 1984, p. 179.
3. *Intense Atmospheric Whirls.* Moscow, Mir, 1985, 368 pp.
4. Maksimov, I.V. *Geophysical Forces and Waters of the Ocean.* Leningrad, Gidrometeizdat,1970, 477pp.
5. Kosty, J. *Large Systems: Connectives, Complexity, and Catastrophes.* M, Mir, 1982
6. Lorenz, E. N. *Nature and Theory of the Atmosphere General Circulation.* Leningrad, Gidrometeizdat, 1970, 259 pp.
7. *Models of the Atmosphere General Circulation.* Leningrad, Gidrometeizdat, 1981, 351 pp.
8. Monin, A. S. *Weather Forecast as a Problem of Physics.* Moscow, Nauka, 1969, 165 pp.
9. *Nonlinear Systems of Hydrodynamic Type.* Moscow, Nauka, 1974, 159 pp.
10. Palmen, E., Newton C. *Circulation Systems of the Atmosphere.* Leningrad, Gidrometeizdat, 1973, 615 pp.
11. Stovas, M. V. *Theory of Critical Parallels and General Circulation of the Atmosphere*, Moscow, Nauka, 1962, 170 pp.
12. Usmanov, R. F. "On the Influence of Rotation on the General Circulation of the Atmosphere.", *Proceedings of CIP.*, 1961, No. 104.
13. Haken, H. *Synergetics: A Hierarchy7 of Instabilities in Self-Organizing Systems.* Moscow, Mir, 1985, 315 pp.
14. Chamberlen, G. *Theory of Planetary Atmospheres.* Moscow, Mir, 1981, 352 pp.
15. Shanina, I. P. *Dynamics of Atmospheric Fronts and Cyclones.* Leningrad, Gidrometeizdat, 1985, 350 pp.

To **Chapter VI**

1. Andronov, A.A., Vit A.A., Khaikin S. E. *Theory of Oscillations*, Moscow, Nauka, 1981, 567 pp.
2. Barenblatt, T. I. *Similarity, Automodelity, Intermediate Asymptotics*. Leningrad, Gidrometeizdat, 1982, 259 pp.
3. Bogolyubov, N.I., Mitropolskii Yu. *A. Asymptotical Methods in the Theory of Nonlinear Oscilations*. Moscow, Nauka, 1974, 503 pp.
4. Blackjoer, O. *Analysis of Nonlinear Systems*. Moscow, Mir, 1969, 210 pp.
5. Dragan, N. I. "General Properties of a Stochastic Model of Rhythnics." *Otbor i Peredacha Informatsii (OPI)*. Kiev, 1975, No.44.
6. Dragan, N.I. "On the Juistification of a Stochastic Model of Rhythmic Phenomena.". *OPI,1971, No.* 34.
7. Marry, J. *Nonlinear Differential Equations in Biology :Lectures on Models,* Moscow, Mir,1983, 397 pp.
8. Mitropolskii, Yu. A., Martinyuk, D. I. *Periodical and Quasi-Periodocal Oscillations of Systems with Delay*. Kiev, " Vitsa shkola ", 1978, 24 pp.
9. Mishchenko, B. O. Rozov I. K., *Differential Equations with Small Parameters and Relaxation Oscilations*. Moscow, Nauka, 1975, 365 pp.
10. Petrovsky, I. G. *Lectures on the Theory of Ordinary Differential Equations*. Moscow, Moscow University Press, 1988, 296 pp.
11. Rabinovich, M. I. Troubetskov, D. I. *Introduction to the Theory of Oscillations and Waves.* Moscow , Nauka, 1984, 429 pp.
12. Sadovsky, M. A." Automodelity of Geodynamic Processes." *Vestnik AN SSSR,* 1986, No. 8
13. Safaryan, N. E. " On Some Characteristics of Geological Cycle Recurrence." .In: *Mathematical Methods of Cycle Recurrence Analyses*. Moscow, Nauka, 1985, 152 pp.
14. Svirezhev, Yu. A. *Nonlinear Waves, Dissipate Structures and Catastrophes in Ecology*. Moscow, Nauka, 1987, 365 pp.
15. Fillipe Ouen, M. "Interaction of Waves." In: Nonlinear Waves. Moscow, Mir, 1977.
16. Khain, V. E. "On General Regularities of Tectonic Process Development in the Course of Time and Problems of Interruption-Noninterruption, Cyclicity- Directness. " *Vestnic MGU, Geol.* 1971, No 4, pp. 3-17
17. Horophas, D. N. *Systems and Simulation*. Moscow, Mir, 1967, 314 pp.

To **Section 6.2**

1. Arnold, V.I. and Avez A. *Ergodic Problems of Classical Mechanics,* N.Y.; Benjamin. 1968.
2. Aihara, K., and others : "Structures of Attractors in Periodically Forced Neural Oscillators." *Pys.Lett. A* 116; 313-17, 1986.
3. Babloyantz, A.,Destexhe A.: "Low Dimensional Chaos in an Instance of Epilepsy." *Proc.Nat Acad Sci.*, USA 83, 3513-17, 1986.
4. Bellman, R. and Dreyfus R. *Applied Dynamic Programming*. Prin.Univ.Press, Princeton, 1961.
5. Benno, Hens, Theodor Plesser : "Temporal and Spatial Order in Biochemical Systems." *Annals of the New York Academy of Sciences* v. 316, Feb.28, 1979.
6. Cindy, L. and others: "Nonlinear Analysis of EEG Sleep Stages." *Neuro-Psychopharmaco-*

logy vol.5, No.3, 1991.
7. Devany, R.L. *An Introduction to Chaotic Systems.* Menlo Park; Benjamin/ Cummings, 1986.
8. Dvorak, I., Siska J. "On some Problems Encountered in the Estimation of the Correlation Dimension of the EEG." , *Phys. Letter.* A 118; 63-66, 1986
9. Eckman, J.P. Rueble D. "Ergodic Theory of Chaos and Strange Attractors."*Rev.Mod.Phys.* 57; 617-56, 1985.
10. Freeman, W.J., Skarda, C.A. "Spatial EEG Patterns, Nonlinear Dynamics and Perception. The Neo Sheringtonianview." *Brain Res.Rev.* 10; 147-75, 1985.
11. Grassberger, P., Procaccia, I. " Characterization of Strange Attractors." .*Phys .Rev. Lett* .50;346-49 12, 1983.
12. Glass, L. Mackey, M. eds. *From Cloaks to Chaos: The Rhythms of Life*. Princeton University Press, Princeton, 1989.
13. Glass, L. Mackey M. "Pathological Conditions Resulting from Instabilities in Physiological Control Systems." *Annals of the N.Y. Academy of Sciences.* v.316, Feb.28, 1979.
14. Halmos P. R. *Lectures on Ergodic Theory.* N.Y. Chelsen, 1958.
15. Halmos P. R. ed. "New Approaches to Nonlinear Problems in Dynamics.". Philadelphia, *PA. Soc. Ind. Appl. Math.* 1980.
16. Hannan E. J. *Multiple Time Series.* N.Y. 1970.
17. Holden, A.V."The Biological Time." , *Priroda,* #6 1966 (in Russian)
18. Jasper, H.H. and others (Eds): *Basic Mechanisms of the Epilepsies*, Little Brown, and Co. 1969
19. Kaczmarek, L.K. Babloyantz, A. " Spatiotemporal Patterns in Epileptics Seizures." Biol.Cyber. 26; 199-208, 1977.
20. Kathleen, M. Crowe. "A Nonlinear Ergodic Theorem for Discrete Systems." *J.Math. Biol.* 32; 179-191,1994
21. Koslow, S.H., and others "Perspective in Biological Dynamics and Theoretical Medicine." *Ann.N.Y. Acad. Sci.* vol.504,1987.
22. Lopes de Silva FH, and others. "Models of Neuronal Population: Three basic mechanisms of rhythicity." In Swab D. Corner MA (eds.) *Progress in Brain Research.* Am sterdam, Else vier, pp281- 308,1976.
23. *Mathematical Modelling of Circadian Systems.* Ed.M.Moore Ede, C.A Czeisler. N.Y.; Reven Press,1984.
24 .Mayer-Kress G. Holzfuss J. "Analysis of the Human Electroencephalogram with Methods from Non-linear Dynamics." In Rensing L., Heiden V.,M Mackey MC (eds). *Temporal order in human Oscillatory Systems*, Berlin, Springer-Verlag, 36;57-68, 1987.
25 .Michel Arbib, David Caplan "Neurolinguistics must be Computational." *The Behavioral and Brain Science,* 2,449-483, 1979.
26. Nunez, PL *Electric Fields of the Brain. The Neurophysics of EEG.* New York, Raven, 1981.
27. Okujava, V. M. *Basic Neurophysiological Mechanisms of Epileptic Activity.* Tbilisi; Gene tleba Puplishing House, 1969.
28. Othmer, H.G. "Applications of Bifurcation Theory in the Analysis of Spatial and Temporal Pattern Formation." . *Annals of the N.Y. Academy of Sciences* v. 316, Feb.28, 1979.
29. Ott, E. "Strange Attractors, Chaotic Motions of Dynamical Systems." *Rev.Mod.Phys.* 53; 655- 671, 1981.
30. Pavlidis, T. *Biological Oscillators; Their Mathematical Analysis.* N.Y. Academic Press, 1973.

31. Sarafyan, N. E. "Some Remarks about the Origin of Chaos in the Cerebrum and its, Possible Causes." *Report. Brain Research Institute*, Harber's Lab. UCLA, 1993.
32. Sarafyan, N. E. *Cybernetic of the Biological Rhythms.* Aiastan Publishing House, 1984. (in Armenian)
33. Sarafyan, N. *Some Aspects of Self-organization.* (in press) 1994
34. Skarda, CA, Freeman, WJ "How Brains Make Chaos in Order to Make Sense of the World." Behave.Brain Sci. 10; 161-95, 1987.
35. Thompson, J.M.T., Stewart M.B., *Non-linear Dynamic and Chaos.* Chichesrter,U.K.; Wiley, 1986.
36. Winfree, A.*The Geometry of Biological Time.* N.Y. Springer-Verlag, 1980.

To **Section 6.3**

1. Ambartsumyan V.A. "*Problems of the Universe's Evolution*". Yerevan , 1968 (in Russian)
2. Arakelyan R.A. "To the Distribution and Interconnection Laws Rectilinear and Curved Forms Reflected on the Cosmic Graph.". *Izv. Vish.Nauch.Zaved.ser.Geology*, 10,1978
3. Balahovsky I.M. "Several Regimes of the Motions Excitation in Idealize Excitable Tissue." *Biophysica,* X,V.6,1965, M. (in Russian)
4. Bazarov A.D. "Hierarchy Moon.". *Vokrug sveta,* #7,1979 (in Russian)
5. Brohakov V.N. "Geology and Cosmos." *Zemlia i Vselenmnaya,* #5, 1979 (in Russian)
6. Burkov V.N. *The Basics of the Mathematical Theory Active Systems,* M. Nauka, 1974 (in Russian)
7. Chechelnitsky A. "Is Solar System Quantized?", *Zemlia i Vselennaya,* #2,1982 (in Russian)
8. *Earth's Continents Ringed Structures*, M. Nauka, 1987 (in Russian)
9. Ebeling V. *Structurbildung die Irreversiblen Processen,* Leipziq, BSB,B.G. Tauber,1976
10. Ebeling W. and Ulbricht H. (eds.) *Self-organization by Nonlinear Irreversible Processes.* Berlin, Springer-Verlag 1986
11. Eigenson M.S. "To the Analogy Solar System and Galaxy Structures.". *Astronomical Journal,* XL, #1,1963 (in Russian)
12. Farrow L.A., Edelson D.M. "The Steady-State Approximation: Fact or Fiction?". *International Journal of Chemical Kinetics*, v.6, 787-800,1974
13. Fedorov K.L. *Physical Nature and Structure Oceanic Fronts*, Leningrad, Gidromedizdat,1983 p. 296 (in Russian)
14. Fedorov K.L."To the Slow Relaxation Thermal Track Hurricane in the Ocean.". *Dokladi AN. USSR,* #4,v.254,1979 (in Russian)
15. Gaponov-Grehov A.V., Rabinovich M.N "Chaotic Dynamic Simplest Systems.". *Priroda* #2, 1981 (in Russian)
16. Gorodetsky E.E. "Convection and Self organization of Structures." *Kvant* #9,1985 (in Russian)
17. Haken G. *Advanced Sinergeties;Instability Hierarchies of Self-organizating Systems and Devices.* Berlin: Springer-Verlag,1983
18. Hess B., Plasser T. "Temporal and Spatial Order in Biochemical Systems.' *Ann. N-Y Academy of Sciences,* v. 316, February 28,1979, p.203
19. *Hydrodynamic Instabilities and the Transition to Turbulence.* By Eds. Swinney H.L.,Gollub J.P New York, Springer, 1981

20. *Instability, Chaos and Predictability in Celestial Mechanics and Stellar Dynamics,* Ed. by K.B Bhatnagar , Nova Science Publishers Inc. 1993
21. *Intensive Atmospheric Vortexes.* Ed. L. Benttonna 1984 (in Russ.)
22. Kadanoff L.P. "Ways to the Chaos." . *Physics Today,* December, 1983, p 46
23. Kerary J. *Orderliness and Disorderliness in the Structures of Material.* Mir, M. 1985 (in Russian)
24. Klafter J., Shlesinger M.J." On the Relationship among Three Theories of Relaxation in Disordered System.". *Proc.Natl.Acad.Sci. USA* v.83, pp848-851, February 1986,Physics
25. Koshliakov M.N. "Synoptical Vortexes of the Open Ocean-Novelty in the Oceanology." *Vesnik AN USSR,*#6,1979 (in Russian)
26. Kratkova O.T. "The Analysis of the Ringed Structures." *Izves.Vuzov Geo.i Razvedka,* #8,1983
27. Krinsky V.I. "Autowavey Processes in the Excitable Media." In *Biophysical Complex Systems and Radiational systems,* Ed. By G.R. Ivanitsky M..Nauka, 1979 (in Russian)
28. Lorenz E.N. *Nature and Theory of the General Circulation of the Atmosphere*, Geneva; Wold Me teorlogical Organization p.312, 1967
29. Maxsimov I.V. *Ocean Waters and Hydrophysic Forces.* Gidrometizdat, L. 1970 (in Russian)
30. Meleketsev I.V. "Vortex Volcanic Hypothesis and Some of its Prospective Applications." In *Depth Problems of the Magmatism*", M. Nauka,1979 (in Russian)
31. Molchanov A.M. "To the Resonance Structure of Solar System.". In . *Modern Problems Celestial Mechanics and Astrodynamics*" M. Nauka, 1973(in Russian)
32. Morowitz H.J. "The Minimum Size of Cells." In *Principles of Bio-molecular Organization* (G.E.W.Wolstenholme, M. O'Conner, eds) pp 446-459, London, 1969
33. Mur-Ed M., Salzmen F. "Temporality Inner Orderliness." In, *Biological Rhythms* Ed. by Aschoff I.,v.1 Plenum Press, New York,1981
34. Nicolis G., Prigogin I. *Self-organization in Non-equilibrium System.* Wiley, New York, 1977
35. *Nonlinear Waves. Self-organization* M. Nauka, 1983, 263 p. (in Russian)
36. Novikov I.V. "Generation Space Structure." *Zemlya i Vselennaya,* # 8, 1981 (in Russian)
37. Ogorodnikov K.F. "Thoughtful to the Earth and Cosmos." *Zemlia i Vselennaya* #5,1980 (in Russian)
38. *Oscillatory Processes in Biological and Chemical Systems,* M. Nauka, 1969, p.440 (in Russian)
39. Odum E. "Fundamentals of Ecology." Philadelphia, Sunders, 1959
40. Odum G. and Odum T. *Men and Nature Energetic Basic,* M. Nauka, 1976
41. Othmer H.G. "Applications of Bifurcation Theory in the Analysis of Spatial and Temporal Pattern Formation." *Ann, N-Y Academy of Sciences,* v.316, February, 28, 1979, p.64
42. Palmen E., Newton R. *Atmospheric Circulation Systems*, New York, Academic Press 1969, p.615
43. Pasha I. "Spiral Galaxy; Development and Idea.", *Zemlia i Vselennaya* #6, 1981 (in Russian)
44. Peebls P.Y. *The Large Scales Structure of the Universe.* Princeton Univ. Press, Princeton 1982
45. Reed V.T. *Dislocation in the Crystals,* M. 1957 (in Russian)
46. *Rheology* Ed. by Arish, M. 1962
47. Romanowsky Y.M. *Two Class of the Self-organization Processes*, Puchino,Institute of Biophysics AN USSR,1982,P.213
48. Sarafyan N.E., Arakelyan R.A. "To the Genesis of Earth Crust Non-linear Structural Formations." Izv.AN . Arm.SSR, Nauki o Zemli, #3,1981 (in Russian)

49. Sarafyan N.E."To the Objective Laws Earth upper Shells Structure." .Izv. AN.Arm.SSR, Nauki o Zemli, #6,1981 (in Russian)
50. Sarafyan N.E. *Cybernetic of the Biorhythms*, Ayastan Puplishing House 1984. (in Armenian)
51. Sarafyan N.E. *Some Problems of Natural Cycles Structures and Interactions*, Ayastan Publishing House 1993. (in Armenian)
52. Sarafyan N.E. "On the Structure of Earth and its Surrounding Space" 1994 (in press)
53. Schmidt-Nelson K. Scaling. *Why is Animal Size so Important*, Cambridge University Press, Cambridge 1984.
54. Scott E. *Waves in the Active and Nonlinear Media, Applied to the Electronics* M. Sov. radio, 1977, 368p. (in Russian)
55. *Self-organization in the Physical, Chemical and Biological Systems* Kichinev, Chtintsa, 1984, p.163 (in Russian)
56. *Self-Organization of the Cybernetic Systems*. Kiev, Institute of Cybernetic, 1980 (in Russian)
57. Shanina I.P. "The Dynamic Atmospheric Fronts and Cyclons.' Leningrad, Gidromedizdat, 1985, p.350
58. Shmalhouzen V.T. *Factors of the Evolution*, Nauka, M. 1984 (in Russian)
59. Sholpo V.N. *Earth's Structure: Orderliness or Disorderliness*, M. Nauka, 1986 p.213 (in Russian)
60. Shoulst S.S. *The Earth Crush Tectonic*, M. Nedra, 1979 (in Russian)
61. Silk D.T. "The Large-Scales Structure of the Universe." In *The World Science*, #12,1983,M (in Russian)
62. *Tides and Resonance in the Solar System, Instability in the Dynamical System*, Ed. by Szebehely V. M. Mir 1975,p.324 (in Russian)
63.Toomre A.B. "Theories of Spiral Structure." *Ann. Rev. Astron. Astrophys. Astron. Astrophys.* 15, 437- 478, 1977
64. Turing A.M. *Phil.Trans. R.Soc.* London, B237,32-72,1952
65. Ugtomsky A.A." Physiological Liability and Balance." In the *Selected Works*, v. VI, Leningrad, 1961
66. Wattington *Organiziers and Genes M.* Nauka, 1945 (in Russian)
67. Winfree A.T. "Spiral Waves of Chemical Activity." , *Science* 175, p.635 - 636, 1972
68. Winner N. Rosenblueth A."The Mathematical Formulation of the Problem of Conduction of Impulses in a Network Connected Excitable Elements, Specifically in Cardiac Muscle.", *Arch.Inst.Cardial. Mex.* 1946,16,p.205-265
69. Yampolski A.D. " Currents in the Ocean." , *Zemlia i Vselennaya*, #1, 1979 (in Russian)
70. Zaikin A.N., Zhabotinsky A.M. " Concentration Waves Propagation in Two dimensional Liquid-phase Self-Oscillatory System" , *Nature* v .225, 1970
71. Zamorsky A.B. "Spiral Motion of Bodies." .*Priroda*, #6, 1965 (in Russian)
72. Zhabotinsky A.M. *Concentrated Self-oscillations*, M. Nauka, 1974 (in Russian)
73. Zikov V.S. *Simulation of the Waves Processes in the Excitable Media* N. Nauka,1984 (in Russian)

To **Section 6.4**

1. Alfven, H. Arrhenius, G. *The Evolution of the Solar System*. San Diego, California, 1976
2. Bazarov, A.D. " Hierarchy Moon." *Vokrug Sveta*, # 7, 1977

3. Bellman, R. *Dynamic Programming*. Princeton University Press, Princeton New Jersey, 1959.
4. Bellman, R. *Adaptive Control Processes: A Guide Tour*. Princeton University Press, Princeton, New Jersey, 1961
5. Binney, J. Tremaine, S. *Galactic Dynamic*. Princeton University Press, Princeton New Jersey, 1987.
6. Chauvin, R. *Les Societes Animales*; De L'Abeille au Gorille, Plon, Paris, 1963
7. Chechelnitsky, A. E. "Is Solar System Quantized." *Zemlia i Vselenaya*, #2, 1982
8. Disney, M. *The Hidden University*. Macmillan, 1984
9. Dressler,A. "The Large-Scale Streaming of Galaxies", *Scientific American*, September,1987.
10. Eigen, M. " Th self-organization of matter and the evolution of biological macromolecules" Naturwissenschaften, 1971, Bd. 58, H, 10, S. 465-523.
11. Eigenson, M.S. "To analogy Solar System and Galaxy Structures." *Astronom.* ***Journal***. XL, 1, Lvov, 1963
12. Feistel, R. Ebeling W Stochastic Models of Evolutionary Processes. In the book "*Termodynamics and Control of Biological Processes*." M. Nauka, 1984.
13. Ferraz-Mello, S. (Ed.) "Chaos, Resonance, and Collective Dynamical Phenomena in the Solar System". *IAU, Symp. 1992*, Kluwer Academic Publishers, Dordrecht Raidel, 1992.
14. Fisher, R.A. *"The genetical theory of natural selection."* Oxford: Dover, 1930.
15. Field, G. "The Hidden Mass in Galaxies." *Mercury*, May/June, 1982
16. Fontana, W. and others "Physical Aspects of Evolutionary Optimization and Adaptation." *Phys. Rev. A. #40, 3301-3321.*
17. Goldrache, P. "A Explanation often Occurring Commensurable Average Motions in the Solar System". In the book *Tides and Resonance in Solar System*, Nauka,M. 1975
18. Goldrache, P. Pil, S. "The Dynamics of the Planets Rotation." In the book *Tides and Resonance in Solar System*, Nauka, M.1975.
19. Haken, H. *Synergetics: An Introduction Nonequilibrium Phase Transitions and Self Organization in Physics, Chemistry and Biology*. Springer-Verlag Berlin Heidelberg New York, 1978
20. Hamilton, W. R. *The Mathematical Papers*. vol. 1, Cambridge, 1931
21. Jeffries, A. and others. "Gravitational Wave Observatories." *Scientific American*, June, 1987
22. Khain, B.E. *The General Geotectonica*, M.Nauka,1973.
23. Lanczos, C. *The Variational Principles of Mechanics*, Toronto, 1962
24. MacDonald, G.J.F. "Tidal Frictions." In the book *Tides and Resonance in Solar System*. Nauka, M. 1975.
25. Melnikov, V.N. "Adaptation as a Multi-level process". In the book, *Adaptation and Adaptagenes*. Vladivostok,1987
26. Monin, S.A., Yaglom A.M. *Statistical Hydromechanics*. M. Nauka, 1965
27. Morrison, L. "Secularly Accelerations Moon's Orbital Movements and Revolution of Earth." In the book *Tides and Resonance in Solar System* , Nauka, M.1975.
28. Nonlinear systems of the hydrodynamics type. Ed. By A.M. Obychov, M. Nauka,1974.
29. Norris, P.W. *Mechanics via the Calculus*, London, 1953.
30. Novikov, I.V. "Generation Space Structure." *Zenlia i Vselenaya*, #8, 1981 30
31. Orbely L.A. *Selected Works v.1* M.1961, pp298-409.

32. Parker, B. " In and around Black Holes." *Astronomy*, October, 1986
33. Polak, L. S. "Variational Principles of Mechanics." In the book *The Variational Principles of Mechanics*, Nauka, M. 1959.
34. Pontriagin L.S., and others. *Mathematical Theory of Optimum Processes*. M. Nauka 1961.
35. Pontriagin, L.S. *Ordinary Differential Equations*. Nauka, M. 1965
36. Romanovsky, Y.M. *Two Class of the Self-Organization Process*. Puchino, Institute of Biophysics AN. SSSR, 1982
37. Rozen, R. *Principles of Optimum in Biology*. M. Nauka, 1969
38. Sarafyan E. N. "Various Principles of Economic System Control." Proceedings Academia of Science of Armenia, Seria of Social Science, #8, 1976.
39. Sarafyan, E. N. About a Hierarchic System Automatons. *Proceedings Academiaof Science of USSR , Tech. Cyb.* # 5, 1968.
40. Shandarin, A.B. and others "Large-Scale Structure of Universe." *UFN*, #3, 1989.
41. Shulman, A.S. and others *Non-stationary Processes*. Minsk, 1986.
42. Silk, D. and others "Large-Scale Structure of Universe." *V Mire Nauki*, #12, 1983.
43. Silk, J. "Formation of the Galaxies." *Sky & Telescope*, December 1986
44. Sussman, G.J. "Chaotic Evolution of the Solar System." *Science 257, 56,1992*.
45. Svirejev, Y. M. and Logofet, D.O. *Stability of Biological Populations*. Nauka, M. 1978
46. Toomre, A.B. "Theories of Spiral Structures." *Ann.Rev.Aston.Astrophys.* Astron.. Astrophys.15, 437-478, 1977.
47. Ugtomsky, A.A. "Physiological Liability and Balance." *Selected Works,* v.VI, Leningrad,1962.
48. Volterra, V. *Lecons sur la Theorie Mathematique de la Lutte Pour La Vie*. Gauther. Paris, 1931
49. Wilson, A .C. "The Molecular Basis of Evolution." *Scientific American*, October, 1985.

To **Section 6.5**

1. *Nonlinear Waves. Self-Organization.* Moscow, 1983, 263 pp.
2. Scott E. *Waves in Active and Nonlinear Media. Application to Electronics*. Moscow, " Sovetskoe Radio ", 1977, 368 pp.
3. Haken G. *Synergetics: A Hierarchy of Instability in Self-Organizing Systems and Devices.* Moscow, Mir, 1985.
4. *Self-Organization of Cybernetic Systems.* Kiev, IK, 1980
5. *Self-organization in Physical, Chemical and Biological Systems*. Kishenev, Shtinitsa, 1984, 163 pp.
6. Ebeling V. *Structure Formation in Irreversible Processes.* Moscow, Mir, 1979, 279 pp.
7. *Oscillating Processes in Biological and Chemical Systems*. Moscow, Nauka, 1957, 440 pp.
8. Sholpo V. I. *Structure of the Earth: Orderliness or Disorderliness*? Moscow, Nauka, 1986.
9. Safaryan N. E. "On the Regularities of the Earth's Upper Shell Structure." *Izv. AN Arm. SSR, Nauki o Zemle*, No. 6, 1981.
10. *Simulation of Hydrophysical Fields and Processes in the Ocean.* Moscow, Nedra, 1986, 255 pp.

11. Lorenz E. I. *Nature and Theory of General Circulation of the Atmosphere*. Leningrad, Gidrometeizdat, 1980, 275 pp.
12. Gorodetsky E. E. "Convection and Structure Self-Organization." *Kvant,* No. 9, 1986.
13. Ugtomsky A. A. "Physical Liability and Balance." In. Selected Works. vol. VI, Leningrad, LGU, 1962.
14. Zamorsky A. V. "Helical Motion of Bodies." *Priroda,* No. 6, 1965.
15. Safaryan N. E. "Perturbation Propagation in Homogeneous with Respect to Relaxation Media." *Izv. AN Arm. SSR, Nauki o zemle*, No. 3, 1984.
16. Nikolis G. Prigozhin I. S. *Self-Organization in Unbalanced Systems*. Moscow, Mir, 1979, 275 pp.
17. Etkins P. *Order and Disorder in the Nature*. Moscow, Mir, 1987
18. Safaryan N. E. "Nature and Self-Organization." , *Gitutyun ev Tekhnika,* No 1, 1989 (in Armenia)
19. Safaryan N. E. *Cybernetics of Biorhythms,* Yerevan 1984, 131 pp. (in Armenia)

To **Section 6.6**

1. Beletsky V. V. *Sketches on Motion of Cosmic Bodies*. Moscow, Nauka, 1977
2. Blekhman I. M. *Synchronization in the Nature and Engineering.* Moscow, Nauka, 1981, 363 pp.
3. Braginskii S. I. "Hydromagnetic Dynamo of the Earth.", *Priroda,* No. 9, 1973, pp. 32-41,
4. Kadomstev B. B., Ridnik V. I. *Waves Around Us.* Moscow, Znamya, 1981, 149 pp.
5. *Components of Adaptative Process.* Ed. V. I. Medvedev, Leningrad, Nauka, 1984, 111 pp.
6. Marsden J., Maccracon M. *Bifurcation of Cycle Origin and its Applications.* Moscow, Mir 1980, 367 pp.
7. Melnicov V. N. "Adaptation as a Multi-Level Process." In: Adaptation and Adaptagneses. Vladivostok, 1977, pp. 21-27
8. Monin S. A. Yaglom A. M. *Statistical Hydromechanics*. Moscow, Nauka, 1965
9. Milanovskii Yu. M. "The periodical Scale of Absolute Age." In: Rhythmicity of Natural Phenomena. Leningrad, Gidrometeizdat, 1971, pp. 66-69
10. Odum Yu. *Fundamentals of Ecology.* Moscow, Mir, 1975, 740 pp.
11. Pankul L. I. *Phoses and Cycles of Planetogenesis.* Alma-Ata, Nauka, 1968, 190 pp.
12. Poushcharovsky Yu. M." Development of Shells and Structures of the Earth." *Priroda,* 1973, No 1, pp. 13-113
14. Rolfs K. *Lectures in Theory of Density Waves.* Moscow, Mir, 1980, 195 pp.
15. Sarkisov D. S. *Sketches on Structural Fundamentals of Homoestasis*. Moscow, Meditsina, 1977, 351 pp.
16. Thompson J. M. T. *Instabilities and Catastrophes in Science and Engineering*. Moscow, Mir, 1985, 254 pp.
17. Tutukov L. V."Stars and Planets." *Nauchnie Informatsii,* No. 63, 1987. Riga.
18. Urlantsev Yu. A. "Specificity of Spatial and Temporal Relations in the Living Nature." In: *Space, Time, Motion.* Moscow, Nauka, 1971 pp. 215-241.
19. Hayasi T. *Nonlinear Oscillations in Physical Systems*. Moscow, Mir, 1968, 11 pp.
20. Hillier P. *Gamma Astronomy.* Moscow, Mir, 1987.
21. Safaryan N. E."Bifurcations and Issues of System's Stability." *Gitutyun ev Tekhnika,* No. 4, 1990, Yerevan (in Armenia)

About the Book

The idea of writing the present book occurred to me for the first time in 1975, when I worked in a program of mathematical simulation of geological processes. Before start of this program, I studied the optimization of finite, probabilistic automatons' collective and multilevel structures' and behaviors', centralization and decentralization of control. My Ph.D. was dedicated to multilevel control systems. The base for such mathematical tasks was the quick processing of information in homogeneous structures, in multilevel structures such as the central nervous system of living beings, the partition of control, and the organization of biological rhythms. Proceeding from these phenomena, it was obvious that the geological, cyclic processes in a life of Earth play the same role as the biological rhythms do in the life-cycle of living beings.

In 1984, I published a book about biological rhythms the main point of which was that the rhythms (their formation and conservation) are extremal mechanisms, which form the equilibrium within cyclic alterations. From this point of view, the geological-tectonic cycles simply are restoration processes permanently arising in the limits of lithosphere, at the interactions with the external forces. The external forces, with respect to the lithosphere, are the Earth's rotational and the mantle's convection forces, which are transmitted continuously to the lithosphere, emerging here tensions and instability. Other geological processes such as earthquakes, volcanic activity, sedimentation, morphological folding are just components of these restoration processes. Their cyclic nature and other characteristics sometimes are interrupted by definite causes. For such conclusions the idea of the reaction's formation, its duration, and composition in the biological active media, has served as a basis. It was the only correct way for the analysis and classification of such intricate problems as were the geological cycles, their geneses, composition, and evolution. Actually, the same issues relate also to the other types of cycles. Inspections of the contradictory hypotheses, about the arising, allocation, and subordination of the linear (lineaments) and nonlinear (spiral-shape) morphological structures in geology, stimulated me again to analyze the existing analogous concept in the non-homogeneous media such as the mechanics of solid bodies, the crystalline structures, the chemical and biological active media. Joint examination of the excitement's spread, the restoration processes', the linear and nonlinear structures' in the non-homogeneous media had testified to their inevitability. Further, the biological rhythms and geological cycles both are optimal mechanisms, which permanently furnish the equilibrium at the changeable external conditions. The structural and functional likeness verifies their irreversibility unity and end.

In 1988, I had handed over for publication my analyses of geological cycles, using the same issues and methods, which we applied in the analyses of the biological rhythms. In the course of time we have started to focus our observation instead of on the question **how**, on the question **why**. The biological rhythmic alterations actually are the continuations of the geological and geophysical cyclic changes. Why? For the further confirmation of these theses, the hydrosphere's and atmosphere's structural and functional contradictory concepts were subjugated to the special considerations. A contradiction, for example, is the genesis of the cyclic circulation in the oceans is connected with the atmospheric cyclic circulation, at the same time, the genesis of the cyclic circulation in the atmosphere is connected with the oceans' cyclic circulation. Our purpose was not so much to make concepts' observation or calculations and get numbers, but more importantly, to determine the structure and direction of transformation (optimal policies), subjugated to various changeable restrictions.

The examinations of the mechanisms of formation, conservation and vanishing of the cyclones (anti-cyclones) as non-linear, of the frontal lines and currents as linear structures, as well as different scales' cyclic movements and circulation confirmed our initial predictions. That is nature is constructed uniformly everywhere, despite all of its diversity. The different scale's linear and non-linear structures are universal and fundamental but the cycles constantly furnish the equilibrium in the shortest time (or minimum dissipation of energy). The general tendency was emphasized evidently everywhere among different scale cyclic changes'. It is just the direction of the energy conversion, in the result the high quality energy emerges. Why? The trigger for all of the transformations (in other words, reactions or restoration processes) is the principle of least actions-**PLA**. From this point of view, the Earth, with its embedded shells, represents an optimum cascade converter of energy, the source of which is external. It is in the solar system, precisely it is compelled from the sun. The core's nuclei, chemical, or gravitational energy is subordinate; they all are just by-products. Such analyses allow us to draw up a conclusive presentation according to which, exogenous geological cycles are basic. Their genesis is linked to the slower, regarding to the Earth's, cycles. The endogenous geological cycles are secondary. Their origin is linked with the faster, regarding to the Earth's, cycles. After such explanation, the center of gravity of study was shifted to the examination and understanding of features of the solar system's cycles: such as revolutionary, rotary and many other cyclic movements of the celestial bodies. Some of the problems' preliminary investigations were summarized in the book *Structural and Interaction some Questions of the Natural Cycles* published in 1992.

Despite the many phenomenal achievements during the XX century, modern astronomy and astrophysics still have numerous unsolved old and new problems and contradictory concepts. For example, the trigger of celestial bodies' hierarchy, the nature of sun cycles', the function of unseen substances, the trigger of radiation, the task of **n** bodies, and so on. Therefore, it seems meaningless to add new uncertainties and attempt to classify and understand them in this area. However, such procedures had many reciprocal opportunities, in which we have been persuaded more than once. The spatial and temporal organization of the geological, geophysical cycles, and biological rhythms, as well as the laws of the formation, conservation, and vanishing of the linear and nonlinear morphological structures in the non-homogenous media created unforeseen pre-requisites. For instance, the trigger of the particles' movements, interrelations, and sizes within the hurricane (cyclone or spiral galaxy) during its lifecycles is only its angular momentum; the inverse movements are determined by compelled moment of momentum; the **n** bodies' movements are subordinated, and "serve" each other; the celestial bodies are quantized not only in mass, but in all ranges of interactions without exclusions.

By them we examined the causes of the self-organization and defined its cardinal trigger. It was just the famous principle of least actions. They yield a clue for the profound understanding of the solar system (and not only the solar system) cycles' origin, their organization, evolution, and interaction. The analyses of analogous ideas in astronomy and astrophysics from the cycles' organization point of view intensified our conclusions. They yield new interpretations and commentary for many problems and extend the limits of our understanding of organization and function of cosmic systems. Some of presentations and approaches are preliminary. These absorbing problems' deep examinations expect to their future researchers. The suggested issues and procedures, even by the first approximation, are constructive in many discussions.

So, this book is not a hodgepodge of concepts and issues. It, on the contrary, represents numerous conscious classifications and decisions (and probably many unconscious synaptic decisions) about cycles, style, word order and sequence, suitability of material, and method of presentation.

About the Author

He graduated with an M.A. from Yerevan State University in mathematics in 1961. After he received his M.A. he spent four years optimizing hydro-technical multistage systems at the Institute of Water Problems. In 1965 he was at Moskow State University and at the Institute of Physiology writing his thesis on multilevel control systems. Simultaneously to his thesis he participated in processing many other programs, such as the optimal partitions of management in economic systems. He obtained his Ph.D. in applied mathematics in 1973. In 1975 he joined the program of mathematical simulation of geological processes at the Institute of Geology, and worked on this program until he moved to the USA. From 1991-1993 he worked at the Brain Institute on the program of the estimation diversity of information (EEG of γ rhythms, registered synchronously by selected channels) according of their attractors.

His research interests are discrete mathematics, the theory of differential equations and probability, theory of optimization and multilevel control systems, their applications in biology, geology, astronomy, and other fields, the systems analyses. He has published numerous research papers, and four books.